STUDENT'S SOLUTIONS MANUAL TO ACCOMPANY

MULTIPLE-CHOICE & FREE-RESPONSE QUESTIONS IN PREPARATION FOR THE AP CALCULUS (BC) EXAMINATION

(EIGHTH EDITION)

By David Lederman

With the assistance of:

Lin McMullin

D&S MARKETING SYSTEMS, INC.
1205 38th Street • Brooklyn, NY 11218

www.dsmarketing.com

ISBN # 1-934780-11-1

PREFACE

This Student's Solution Manual is a supplement to MULTIPLE-CHOICE & FREE-RESPONSE QUESTIONS IN PREPARTION FOR THE AP CALCULUS (BC) EXAMINATION, EIGHTH EDITION.

This manual is a valuable supplement for those students who desire to see the steps involved in working out the problems. The student should try to become independent of this manual. Only use it after you have tried a problem and are "stuck".

I will be delighted to hear from students and teachers who come up with neat and elegant ways of solving the problems in this workbook. I would also appreciate hearing from you if you find any errors, typographical or mathematical.

I wish to acknowledge several people whose help and encouragement were invaluable in the production of this manual. I wish to thank Bob Byrne of St. Thomas Aquinas HS (Fort Lauderdale, FL) for sharing his ideas and making valuable suggestions for some solutions which have been incorporated in this edition. A special thanks to Lin McMullin for submitting solutions to the questions which he submitted for this edition. I wish to thank Tammi Pruzansky of D&S Marketing Systems, Inc., for preparing the graphs and typesetting the entire manuscript. Finally, I am grateful to my wife for her encouragement and understanding of the many hours required in the production of this manual.

Any errors found in this book are solely the responsibility of the author.

All communications concerning this book should be addressed to:

D & S Marketing Systems, Inc.
1205 38th Street
Brooklyn, NY 11218
www.dsmarketing.com

TABLE OF CONTENTS

Sample Examination I

1. To be continuous both parts of the function must have the same value at $x = 5$.

$$5^2 + 5b = 5\sin\left(\frac{\pi}{2}(5)\right)$$
$$25 + 5b = 5(1)$$
$$5b = -20$$
$$b = -4$$

The correct choice is (C).

2. $y = 3x^2 - x^3$

$y' = 6x - 3x^2 = 3x(2 - x)$

The critical points are at $x = 0$ and $x = 2$.

Using the First Derivative Test to find possible relative max/min,

$$y' \quad \begin{array}{c|c|c} x < 0 & 0 < x < 2 & x > 2 \\ \hline - & + & - \end{array}$$

$x = 2$ is the only relative maximum (where y' switches from + to –).

Therefore, $(2,4)$ is the only relative maximum of $y = 3x^2 - x^3$.

The correct choice is (C).

3. $v(t) = \langle t^2, \sin(\pi t) \rangle$

$$s(t) = \left\langle \frac{t^3}{3} + C, -\frac{1}{\pi}\cos(\pi t) + K \right\rangle$$

Since $s(0) = \langle 1, 0 \rangle$ from the given information,

$$s(t) = \left\langle \frac{t^3}{3} + 1, -\frac{1}{\pi}\cos(\pi t) + \frac{1}{\pi} \right\rangle, \text{ and}$$

$$s(3) = \left\langle 10, -\frac{1}{\pi}\cos(3\pi) + \frac{1}{\pi} \right\rangle = \left\langle 10, \frac{2}{\pi} \right\rangle$$

The correct choice is (B).

1

4. Use implicit differentiation to find $\dfrac{dy}{dx}$:

$$2y\frac{dy}{dx} - 3x^2 - 30x = 0$$

$$2y\frac{dy}{dx} = 3x^2 + 30x$$

$$\frac{dy}{dx} = \frac{3x(x+10)}{2y}$$

The derivative is zero when $x = -10$ and when $x = 0$. These are the values of x for which there are horizontal tangents.

The derivative may also be found by solving for y and then differentiationg:

$$y = \pm\sqrt{x^3 + 15x^2 + 4}$$

$$\frac{dy}{dx} = \frac{3x^2 + 30x}{\pm 2\sqrt{x^3 + 15x^2 + 4}}$$

$$= \frac{3x(x+10)}{\pm 2\sqrt{x^3 + 15x^2 + 4}}$$

Again when $x = -10$ and $x = 0$, the curve will have horizontal tangent lines.

The correct choice is (D).

5. This geometric series will converge when $|3x - 5| < 1$. Dividing this absolute value inequality by 3 yields $\left|x - \frac{5}{3}\right| < \frac{1}{3}$. Thus the radius of convergence is $\frac{1}{3}$.

The correct choice is (A).

6. The given expression is of the form $\dfrac{0}{0}$, so use L'Hôpital's Rule to find the required limit. The derivative of the integral is found by using the Fundamental Theorem of Calculus.

$$\lim_{x \to 0} \frac{\displaystyle\int_1^{1+x} \frac{\cos t}{t}\,dt}{x} = \lim_{x \to 0} \frac{\frac{\cos(1+x)}{(1+x)}}{1} = \cos 1$$

The correct choice is (B).

7. $f(x) = \sqrt{4 \sin x + 2} = (4 \sin x + 2)^{\frac{1}{2}}$

$f'(x) = \frac{1}{2}(4 \sin x + 2)^{-\frac{1}{2}}(4 \cos x) = \frac{2 \cos x}{\sqrt{4 \sin x + 2}}$

Therefore, $f'(0) = \frac{2 \cos(0)}{\sqrt{4 \sin(0) + 2}} = \frac{2(1)}{\sqrt{0 + 2}} = \frac{2}{\sqrt{2}}$ or $\sqrt{2}$

The correct choice is (E).

8. Consider a thin strip of the forest x_i miles from the longer edge. This strip has an area of $5\Delta x$ square miles. The density gives the number of trees per square mile, so the number of trees in this strip is $\rho(x_i)(5\Delta x)$. The sum of all such strips is $\sum_{i=1}^{n} \rho(x_i)(5\Delta x)$. The limit of this Riemann sum is $5\int_{0}^{3} \rho(x)\, dx$ and this gives the total number of trees in the forest.

The correct choice is (C).

9. By implicit differentiation of $x^2 + y^2 = 169$, $2x + 2y \frac{dy}{dx} = 0$

Solve for $\frac{dy}{dx}$: $\frac{dy}{dx} = \frac{-x}{y}\Big|_{(5,-12)} = \frac{5}{12}$

The tangent line to the circle $x^2 + y^2 = 169$ at $(5, -12)$ has a slope of $\frac{5}{12}$.

The equation of the tangent line is $y - y_1 = m(x - x_1)$, or

$$y - (-12) = \frac{5}{12}(x - 5) \text{ or } y + 12 = \frac{5}{12}(x - 5).$$

Multiply both sides of the equation by 12 and simplify, thus $5x - 12y = 169$.

The correct choice is (C).

10. The speed is the absolute value of the velocity. The speed is the greatest when the velocity is farthest from zero (above or below). This occurs at point D when the (negative) velocity has a larger absolute value than B where the velocity is greatest.

The correct choice is (D).

11. Applying the ratio test to the given series,

$$\lim_{n \to \infty} \left| \frac{(-1)^{n+1} x^{n+1}}{\ln(n+1)} \div \frac{(-1)^n x^n}{\ln n} \right| = \lim_{n \to \infty} \frac{\ln(n)}{\ln(n+1)} |x| = |x| < 1.$$

Therefore, the series converges absolutely if $|x| < 1$. Now test the endpoints.

If $x = 1$, the series $\sum_{n=2}^{\infty} \frac{(-1)^n}{\ln n}$ converges by the Alternating Series Test.

If $x = -1$, the series $\sum_{n=2}^{\infty} \left(\frac{1}{\ln n} \right)$ diverges by the direct comparison test with the harmonic series,

i.e., $\sum_{n=2}^{\infty} \frac{1}{\ln n} > \sum_{n=2}^{\infty} \frac{1}{n}$, which diverges.

Therefore, the interval of convergence is $-1 < x \le 1$.

The correct choice is (D).

12. Recall that $\int \frac{du}{\sqrt{a^2 - u^2}} = \sin^{-1}\left(\frac{u}{a}\right) + C$

Let $u = x$ and $a = 2$ with $du = dx$.

Thus $\int \frac{dx}{\sqrt{4 - x^2}} = \int \frac{du}{\sqrt{a^2 - u^2}} = \sin^{-1}\left(\frac{u}{a}\right) + C$ or $\sin^{-1}\left(\frac{x}{2}\right) + C$

Therefore, the correct choice is (A).

13. In general, a function $f(x)$ has a point of inflection where $f''(x) = 0$ or when $f''(x)$ is undefined, and the concavity changes at that point.

In the given problem, $f'(x) = 4x - \frac{k}{x^2}$ and $f''(x) = 4 + \frac{2k}{x^3}$.

Since $f(x)$ has a point of inflection at $x = -1$, then $4 + \frac{2k}{(-1)^3} = 4 - 2k = 0$

or $k = 2$.

The correct choice is (E).

14. Using integration by parts, $u = x \quad dv = \sec^2 x \, dx$
$$du = dx \quad v = \tan x$$

$$\int x \sec^2 x \, dx = x \tan x - \int \tan x \, dx = x \tan x - (-\ln|\cos x|) + C$$

$$= x \tan x + \ln|\cos x| + C$$

Therefore, $f(x) = x \tan x$.

The correct choice is (C).

15. $\dfrac{dx}{dt} = -3e^{-t}$ and $\dfrac{dy}{dt} = 6e^t$

$$\frac{dy}{dx} = \frac{\frac{dy}{dt}}{\frac{dx}{dt}} = \frac{6e^t}{-3e^{-t}}\bigg|_{t=0} = \frac{6}{-3} \text{ or } -2$$

When $t = 0$, $x = 3e^0$ or $x = 3$, and $y = 6e^0$ or $y = 6$.

The equation of the tangent line is $y - 6 = -2(x - 3) = -2x + 6$ or $2x + y - 12 = 0$

The correct choice is (A).

16. By partial fractions,

$$\frac{1}{2x^2 + 3x + 1} = \frac{A}{2x + 1} + \frac{B}{x + 1}$$
$$1 = A(x + 1) + B(2x + 1)$$

If $x = -1$, $1 = -B \Rightarrow B = -1$.

If $x = -\dfrac{1}{2}$, $1 = \dfrac{1}{2}A \Rightarrow A = 2$.

Therefore, $\dfrac{1}{2x^2 + 3x + 1} = \dfrac{2}{2x + 1} - \dfrac{1}{x + 1}$

$$\int \frac{dx}{2x^2 + 3x + 1} = \int \frac{2}{2x + 1} \, dx - \int \frac{dx}{x + 1} = \ln|2x + 1| - \ln|x + 1| + C = \ln\left|\frac{2x + 1}{x + 1}\right| + C$$

The correct choice is (D).

17. $\dfrac{dx}{dt} = \cos t$ and $\dfrac{dy}{dt} = -2 \sin t \cos t$

$$\frac{dy}{dx} = \frac{\frac{dy}{dt}}{\frac{dx}{dt}} = \frac{-2 \sin t \cos t}{\cos t} = -2 \sin t$$

$$\frac{d^2 y}{dx^2} = \frac{\frac{d}{dt}\left(\frac{dy}{dx}\right)}{\frac{dx}{dt}} = \frac{-2 \cos t}{\cos t} = -2$$

Therefore, $\dfrac{d^2 y}{dx^2}$ is always -2.

The correct choice is (A).

18. Carefully translate the sentence into symbols.

The rate of change of the surface area of a cube, S, with respect to time, t, is $\dfrac{dS}{dt}$, which is directly proportional (k times) to the square root of one-sixth of the surface area; $k\sqrt{\dfrac{S}{6}}$.

The correct choice is (C).

19. Refering to the accompanying diagram,

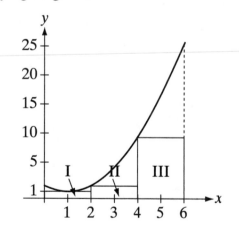

Area of Rectangle I: $2 \cdot f(1) = 2 \cdot 1 = 2$
Area of Rectangle II: $2 \cdot f(2) = 2 \cdot 2 = 4$
Area of Rectangle III: $2 \cdot f(4) = 2 \cdot 10 = 20$
Total Area of Rectangles: $2 + 4 + 20 = 26$

The correct choice is (B).

20. Notice that the segments that make up the slope field are always parallel across the page. This means that the x-coordinates do not affect the slope and therefore x does not appear in the differential equation. This eliminates choices (A), (B) and (C). The slope field for choice (D) can contain only segments with positive slopes. This eliminates (D).

The correct choice is (E).

21. The series is geometric with a common ratio of $2x$. It will converge when

$$|2x| < 1$$

$$-1 < 2x < 1$$

$$-\frac{1}{2} < x < \frac{1}{2}$$

The correct choice is (B).

22. The well known series for $\sin(x)$ is $x - \frac{x^3}{3!} + \frac{x^5}{5!} - \frac{x^7}{7!} + \cdots$

Substitute x^2 for x to obtain the series for

$$\sin(x^2) = x^2 - \frac{(x^2)^3}{3!} + \frac{(x^2)^5}{5!} - \frac{(x^2)^7}{7!} + \cdots = x^2 - \frac{x^6}{3!} + \frac{x^{10}}{5!} - \frac{x^{14}}{7!} + \cdots$$

Then divide each term by x^2 to obtain the required series:

$$\frac{\sin(x^2)}{x^2} = 1 - \frac{x^4}{3!} + \frac{x^8}{5!} - \frac{x^{12}}{7!} + \cdots = \sum_{k=0}^{\infty} \frac{(-1)^k x^{4k}}{(2k+1)!}$$

You may also work directly from the summation notation:

$$\sin(x) \sum_{k=0}^{\infty} \frac{(-1)^k x^{2k+1}}{(2k+1)!}$$

$$\sin(x^2) \sum_{k=0}^{\infty} \frac{(-1)^k (x^2)^{2k+1}}{(2k+1)!} = \sum_{k=0}^{\infty} \frac{(-1)^k x^{4k+2}}{(2k+1)!}$$

$$\frac{\sin(x^2)}{x^2} = \frac{1}{x^2} \sum_{k=0}^{\infty} \frac{(-1)^k x^{4k+2}}{(2k+1)!} = \sum_{k=0}^{\infty} \frac{(-1)^k x^{4k}}{(2k+1)!}$$

The correct choice is (D).

23. The length of the curve $y = f(x)$ from $x = a$ to $x = b$ is $L = \int_a^b \sqrt{1 + [f'(x)]^2}\, dx$.

Therefore, $1 + [f'(x)]^2 = e^{2x} + 2e^x + 2$

$$[f'(x)]^2 = e^{2x} + 2e^x + 1 = (e^x + 1)^2$$
$$f'(x) = e^x + 1$$

To find $f(x)$, integrate $f'(x)$: $f(x) = \int (e^x + 1)\, dx = e^x + x + C$.

So any function of the form $f(x) = e^x + x + C$, where C is any constant, will have a length of $\int_a^b \sqrt{e^{2x} + 2e^x + 2}\, dx$.

The only choice of this form is $e^x + x - 2$.

The correct choice is (E).

24. By the Product Rule $h'(x) = f(x)\, g'(x) + g(x)\, f'(x)$ and

$$h'(3) = f(3)\, g'(3) + g(3)\, f'(3) = (1)(1) + (3)\left(-\frac{1}{3}\right) = 0$$

The values of the functions are read from the graph; the values of the derivatives are the slopes of the lines at $x = 3$.

The correct choice is (C).

25. $C = 2\pi r$ and therefore $\dfrac{dC}{dt} = 2\pi \dfrac{dr}{dt}$. Since $\dfrac{dC}{dt} = 0.5$, $\dfrac{dr}{dt} = \dfrac{0.5}{2\pi} = \dfrac{1}{4\pi}$ meters/minute.

$A = \pi r^2$ and $\dfrac{dA}{dt} = 2\pi r \dfrac{dr}{dt}$. When $r = 4$ meters, $\dfrac{dA}{dt} = 2\pi (4)\dfrac{1}{4\pi} = 2\, \text{m}^2/\text{min}$.

The correct choice is (A).

26. The integral $\int_1^4 \dfrac{t-2}{(t+1)(t-4)}\, dt$ is an improper integral, since the integrand has a vertical asymptote at $t = 4$. Since the limits of integration are $t = 1$ and $t = 4$, you must rewrite $\int_1^4 \dfrac{t-2}{(t+1)(t-4)}\, dt$ as $\displaystyle\lim_{b \to 4^-} \int_1^b \dfrac{t-2}{(t+1)(t-4)}\, dt$.

The correct choice is (E).

27. The average value is $\dfrac{1}{0-(-4)}\displaystyle\int_{-4}^{0}\cos\left(\tfrac{1}{2}x\right)dx = \tfrac{1}{4}\displaystyle\int_{-4}^{0}\cos\left(\tfrac{1}{2}x\right)dx.$

Let $u = \tfrac{1}{2}x,\ du = \tfrac{1}{2}dx \Rightarrow dx = 2\,du$

Since $\displaystyle\int\cos\left(\tfrac{1}{2}x\right)dx = 2\int\cos(u)\,du = 2\sin\left(\tfrac{1}{2}x\right)+C$

Then, $\tfrac{1}{4}\displaystyle\int_{-4}^{0}\cos\left(\tfrac{1}{2}x\right)dx = \tfrac{1}{4}\cdot 2\sin\left(\tfrac{1}{2}x\right)\Big|_{-4}^{0} = \tfrac{1}{2}\sin\left(\tfrac{1}{2}x\right)\Big|_{-4}^{0} = \tfrac{1}{2}\left(0-\sin(-2)\right) = -\tfrac{1}{2}\sin(-2)$

Since $\sin(-2) = -\sin(2)$, then $-\tfrac{1}{2}\sin(-2) = -\tfrac{1}{2}(-\sin(2)) = \tfrac{1}{2}\sin(2).$

The correct choice is (E).

28. If $u = \sqrt{2x}$, then $du = \dfrac{2}{2\sqrt{2x}}dx$. Thus, $du = \dfrac{dx}{u}$ and $dx = u\,du.$

Adjusting the limits of integration, when $x = 2, u = 2$ and when $x = 8, u = 4$. Making all these

substitutions $\displaystyle\int_{2}^{8}\dfrac{dx}{\sqrt{2x}+1} = \int_{2}^{4}\dfrac{u\,du}{u+1}.$

The correct choice is (D).

29. The graph of the region described is:

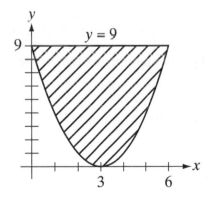

Using the disk method, the radius is $9 - (x-3)^2$ and the volume equals

$$\pi\int_{0}^{6}\left(9 - (x-3)^2\right)^2 dx$$

Since the parabola $y = (x-3)^2$ is symmetric about the line $x = 3$, the volume could then be

given as: $2\pi\displaystyle\int_{0}^{3}\left(9 - (x-3)^2\right)^2 dx$

The correct choice is (B).

30. The tangent lines will be parallel when the derivatives of f and g are equal.

$$f'(x) = g'(x)$$
$$\sec^2 x = 2x$$
$$\frac{1}{\cos^2(x)} = 2x$$
$$x \approx 2.083$$

Use the solver feature on a calculator or find the intersection of the two functions..

The correct choice is (C).

31. The average rate of change of f is given by $\dfrac{\Delta f}{\Delta t} = \dfrac{\frac{1}{b} - \frac{1}{a}}{b - a} = \dfrac{a - b}{ab(b - a)} = -\dfrac{1}{ab}$. The derivative

is $f'(t) = -\dfrac{1}{t^2}$. Set these equal to each other and solve for t:

$$-\frac{1}{ab} = -\frac{1}{t^2}$$
$$t^2 = ab$$
$$t = \sqrt{ab}$$

The correct choice is (B).

32. $f'(x) = 0 + \cos(x^2) = 0$

$$x^2 = \frac{\pi}{2} + k\pi$$

$$x = \sqrt{\frac{\pi}{2}} \text{ is the smallest positive value}$$

$$x = 1.253$$

The equation may also be solved using a graphing calculator.

The correct choice is (B).

33. $R(t)$ is the rate of change of the volume of the water and therefore a derivative. Thus by the Fundamental Theorem of Calculus, $\displaystyle\int_0^3 R(t)\, dt$ represents the total amount of water that leaks out in the first three hours.

Keep in mind that the definite integral of a rate of change gives total change; the integral of the rate at which water leaks is the amount of water that leaked.

The correct choice is (C).

34.

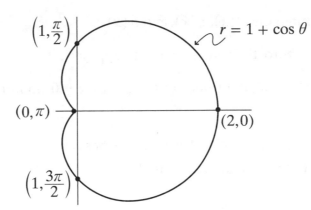

The polar curve $r = 1 + \cos \theta$ is a cardioid which is symmetric about the x-axis as shown in the figure above.

The area is $A = \dfrac{1}{2}\displaystyle\int_0^{2\pi}(1 + \cos \theta)^2\,d\theta$.

This is not one of the choices; however, since $r = 1 + \cos \theta$ is symmetric about the x-axis, the area can be found by doubling the area above the x-axis, and can be written as

$$2\left(\frac{1}{2}\int_0^{\pi}(1 + \cos \theta)^2\,d\theta\right) \text{ or the area } A = \int_0^{\pi}(1 + \cos \theta)^2\,d\theta.$$

The correct choice is (B).

35. Since f is increasing, $\dfrac{f(b) - f(0)}{b - 0} > 0$. Since this is the value of $f'(c)$, $f'(c) > 0$. (D) is true and (C) is false.

(A) is false: $f''(c) < 0$ may be true, but only if the function is concave down. (B) is false since it assumes $f(0) = 0$. (E) is false since this implies that the function is not always increasing.

The correct choice is (D).

36. The expression $\pi(R(x))^2$ is the cross section area of the pipe. Since x measures the position at which the cross section is measured, integrating in terms of x from $x = 10,000$ to $x = 30,000$ gives the volume of that section of the pipe, not necessarily the amount of water flowing through the pipe.

The correct choice is (B).

37. Examine each of the functions $f(x)$, $g(x)$, and $h(x)$.

$f(x)$: As x approaches zero, the values appear to approach 2.

$g(x)$: Even though $g(0) = 2$, the values to the left of $x = 0$ do not approach 2, so 2 is not the limit.

$h(x)$: The values of h are approaching 2 as x approaches zero. The fact that the function is not defined at $x = 0$ does not affect the limit.

The correct choice is (D).

38. Solve the differential equation by separating the variables:

$$y\,dy = x^2 dx$$

Antidifferentiate both sides:

$$\frac{1}{2}y^2 = \frac{1}{3}x^3 + C$$

Substitute the initial condition ($x = 1$ when $y = 1$) to find the constant C:

$$\frac{1}{2}(1) = \frac{1}{3}(1) + C \Rightarrow C = \frac{1}{6}$$

$$\frac{1}{2}y^2 = \frac{1}{3}x^3 + \frac{1}{6}$$

$$y^2 = \frac{2}{3}x^3 + \frac{1}{3}$$

$$y = \sqrt{\frac{2x^3 + 1}{3}}$$

The correct choice is (E).

39. The local maximums and minimums of $f'(x)$ correspond to points of inflection on the graph of function $f(x)$. For example, at the top left maximum point, $f'(x)$ changes from increasing to decreasing, thus the function $f(x)$ changes from concave up to concave down. Another way to find the point of inflection is that at the extreme points of $f'(x)$, the derivative's derivative of $f(x)$, the second derivative of $f(x)$ changes sign, indicating a point of inflection. There are four such points shown in the graph of $f'(x)$.

The correct choice is (E).

40. The average value of a function on an interval $[a,b]$ is given by $\dfrac{1}{b-a}\displaystyle\int_a^b f(x)\,dx$. The integral gives the area under the graph. So in this problem the average value is

$$\frac{\text{area}}{12} = \frac{4 \cdot 8 + \frac{1}{4}\pi(4^2)}{12} \approx 3.714.$$

The correct choice is (B).

41. The third-degree Maclaurin polynomial for the $\sin(x)$ is $x - \dfrac{x^3}{3!}$ (this is one of the power series that should be memorized). Graph this and $y = x^2$ in the same window (see Figure 1). The graph shows 2 points of intersection, BUT there is another point. The graph is a cubic function and a quadratic function. <u>Cubic functions grow faster than quadratics</u>. There is a third point of intersection in the second quadrant as figure 2 shows. It should *not* be necessary to produce the second graph to know that they *will* intersect again.

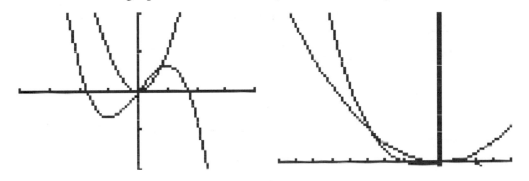

Figure 1
Window is $[-4,4]$ by $[-2,2]$

Figure 2
Window is $[-9,4]$ by $[-2,70]$

The correct choice is (D).

42. The rate of production is given by the first derivative of A: $A'(t) = 48 - 12(t-3)^2$.

The maximum increase in the rate of production occurs when this expression has its maximum value, that is, when $A''(t) = 0$.

$A''(t) = -24(t-3) \Rightarrow A''(t) = 0$ when $t = 3$. Three hours after 8:00 am is 11:00 am.

<u>Note</u>: $A'''(3) = -24$ indicating that $t = 3$ is a maximum (Second Derivative Test). Or use the First Derivative Test: A'' switches from positive to negative at $t = 3$, so $t = 3$ is a maximum.

The correct choice is (C).

43. Using the Product Rule the derivative is

$$4x^2(-\sin(x)) + (\cos(x))(8x) = 8x\cos(x) - 4x^2\sin(x).$$

The correct choice is (B).

44. The limiting value is 6000 as time goes on, i.e. $\lim\limits_{t \to +\infty}(6000 - 5500e^{-0.159t}) = 6000$. Using a calculator, find the intersection of $P(t)$ and $P = 3000$. Another possibility is to solve the equation

$$3000 = 6000 - 5500e^{-0.159t}$$
$$-3000 = -5500e^{-0.159t}$$
$$\frac{30}{55} = e^{-0.159t}$$

$$\ln\left(\frac{30}{55}\right) = -0.159t \Rightarrow t = \frac{\ln\left(\frac{30}{55}\right)}{-0.159} \Rightarrow t \approx 3.8$$

Thus the population will reach half of its limiting value in the <u>fourth</u> year.

The correct choice is (C).

45. Since $f'(x) > 0$ for $0 < x < 6$; and $f'(x) < 0$ for $x > 6$, $f(x)$ increases for $0 < x < 6$ and $f(x)$ decreases for $x > 6$. Obviously, the maximum point on the graph of $f(x)$ is at $x = 6$.

Therefore, the maximum value of $f(x)$ is $f(6)$. To find $f(6)$, use the Fundamental Theorem of Calculus:

$$\int_2^6 f'(x)\,dx = f(6) - f(2) \text{ or, } f(6) = f(2) + \int_2^6 f'(x)\,dx$$

Since it is given that $f(2) = 10$, then $f(6) = 10 + \int_2^6 f'(x)\,dx$.

Now, $\int_2^6 f'(x)\,dx$ is the area under $f'(x)$ from $x = 2$ to $x = 6$, which is approximately 100. Note that each square box has an area of 5, and there are approximately 20 square boxes combined.

Therefore, the maximum value of $f(x) \approx 10 + 100 \approx 110$.

The correct choice is (E).

1a. At any point the slope of the hill is given by $\dfrac{d}{dx}f(x) = -\dfrac{50}{100}\sin\left(\dfrac{x}{100}\right)$. An equation of the tangent line at $x = a$ is:

$$y - 50\cos\left(\dfrac{a}{100}\right) = -\dfrac{50}{100}\left(\sin\left(\dfrac{a}{100}\right)\right)(x - a)$$

$$y = -\dfrac{1}{2}\left(\sin\left(\dfrac{a}{100}\right)\right)x + \dfrac{a}{2}\sin\left(\dfrac{a}{100}\right) + 50\cos\left(\dfrac{a}{100}\right)$$

1b. The top of the hill is at $(0,50)$ and the person's eyes are at $(0,55)$, which is the y-intercept of the line. Therefore,

$$\dfrac{a}{2}\sin\left(\dfrac{a}{100}\right) + 50\cos\left(\dfrac{a}{100}\right) = 55$$

Solve this equation on your calculator for values in the approximate interval $[0,100\pi]$ covering the top to the bottom of the hill. $\underline{a = 45.935}$. (The tangent also intersects the hill at $a = 398.341$, but this is obviously past the middle of the valley.)

1c. The middle of the valley is at $x = 100\pi$ (half the period of $f(x) = 50\cos\left(\dfrac{x}{100}\right)$).

$f(100\pi) = -50$ while $y(100\pi) = -\dfrac{1}{2}\left(\sin\left(\dfrac{45.935}{100}\right)\right)(100\pi) + 55 = -14.643$.

Since $25 < -14.643 - (-50)$ the top of the flagpole is at $(100\pi\quad 25)$, which is below the person's line of sight along the tangent line. Hence the top of the flagpole cannot be seen by the person.

2a.

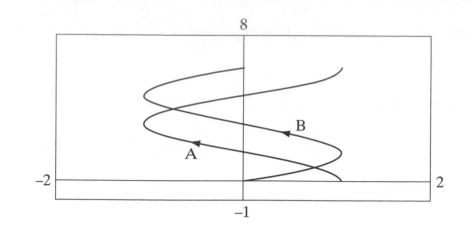

2b. The velocity vector is the derivative of the position vector:

For A: $\langle -\sin t, 1 \rangle$

For B: $\langle \cos t, 1 \rangle$

2c. At $t = 5$, the x-component of the velocity vector for A is $-\sin(5) = 0.959$ and for B it is $\cos(5) = 0.284$. Therefore, A is moving to the right faster.

2d. For all t, including $t = 5$, the y-component of the velocity vector of both particles is the same (i.e. they are both 1) – therefore, both particles are moving upwards at the same rate.

REMINDER: Numeric and algebraic answers need not be simplified, and if simplified incorrectly credit will be lost.

3a. $x(3) = 7(3) - 4(3^2) + \int_0^3 s^2 \, ds = 21 - 36 + \left.\frac{s^3}{3}\right|_0^3 = 21 - 36 + 9 = -6$

$v(t) = x'(t) = 7 - 8t + t^2$

$v(3) = 7 - 8(3) + 3^2 = -8$

3b. Speed $= |v(3)| = 8$.

$a(t) = v'(t) = -8 + 2t$

$a(3) = -2$

Since the velocity and acceleration have different signs, the speed is decreasing.

3c. $v(t) = 7 - 8t + t^2 = (t - 7)(t - 1) = 0$

For $0 < t < 1, v(t) < 0$, and for $1 < t < 4, v(t) > 0$. Hence there is a change in direction of the particle at $t = 1$.

(Note: Although $v(7) = 0, t = 7$ is outside of the given domain and therefore not considered as a solution value.)

3d. Since the particle changes direction only once, check the function's value at that point and also at the endpoints of the given domain.

$x(0) = 0$

$x(1) = 7(1) - 4(1^2) + \int_0^1 s^2 \, ds = 7 - 4 + \left.\frac{s^3}{3}\right|_0^1 = 7 - 4 + \frac{1}{3} = \frac{10}{3}$

$x(4) = 7(4) - 4(4^2) + \int_0^4 s^2 \, ds = 7 - 4 + \left.\frac{s^3}{3}\right|_0^4 = 28 - 64 + \frac{64}{3} = -\frac{44}{3}$

The particle is farthest right at $t = 1$ and farthest left at $t = 4$.

4a. $\int_1^\infty \frac{1}{x^2}\,dx = \lim_{b\to\infty} \int_1^b \frac{1}{x^2}\,dx = \lim_{b\to\infty}\left(-\frac{1}{x}\right)\Big|_1^b = \lim_{b\to\infty}\left(-\frac{1}{b}-(-1)\right) = 1$

The integral converges to 1.

4b. $\int_1^\infty \left(1-\frac{1}{x^2}\right)dx = \lim_{b\to\infty} \int_1^b \left(1-\frac{1}{x^2}\right)dx = \lim_{b\to\infty}\left(x+\frac{1}{x}\right)\Big|_1^b = \lim_{b\to\infty}\left(\left(b+\frac{1}{b}\right)-(1+1)\right) = \infty$

The integral diverges.

4c. $\int_1^\infty \left(\frac{1}{x^2}\right)^2 dx = \lim_{b\to\infty} \int_1^b \frac{1}{x^4}\,dx = \lim_{b\to\infty}\left(-\frac{1}{3x^3}\right)\Big|_1^b = \lim_{b\to\infty}\left(-\frac{1}{3b^3}-\left(-\frac{1}{3}\right)\right) = \frac{1}{3}$

The integral converges to $\frac{1}{3}$.

5a.

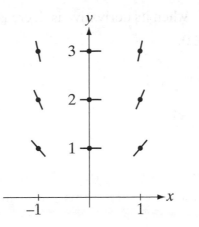

5b. Solve the differential equation by separating the variables and integrating both sides.

$$y^{-2} dy = x \, dx$$

$$\frac{-1}{y} = \frac{x^2}{2} + C$$

$$y = \frac{-1}{\frac{x^2}{2} + C}$$

5c. Substitute the initial condition:

$$\frac{-1}{0^2 + C} = 1$$

$$C = -1$$

$$y = \frac{-1}{\frac{x^2}{2} - 1} = \frac{-2}{x^2 - 2}$$

5d. The functions will have vertical asymptotes when the denominator has real roots. This will occur when $C \le 0$.

6a. The function, f, is concave down when its derivative is decreasing. This occurs on the interval $[-1, 1.5]$. On this interval, $G'' \leq 0$.

<u>Note</u>: $G'(x) = f(x)$ and $G''(x) = f'(x)$

Since $G'(x) = \dfrac{d}{dx} \displaystyle\int_{-2}^{x} f(t)\, dt = f(x)$

6b. $m = 2$, $G(0) = 7$. From the graph, $G'(0) = f(0) = 2$ is the slope of the tangent line. The point $(0, 7)$ is on the tangent line at $x = 0$, so this point must also be on the graph of G. Therefore $G(0) = 7$.

6c. Since the average value of f on $0 \leq x \leq 3$ is $\dfrac{1}{3-0} \displaystyle\int_0^3 f(x)\, dx = 0$, $\displaystyle\int_0^3 f(x)\, dx = 0$, and

$G(0) = 7$

$G(3) = G(0) + \displaystyle\int_0^3 f(x)\, dx$

$\qquad = 7 + 0$

$\qquad = 7$

<u>Note</u>: $G(3) = G(-2) + \displaystyle\int_{-2}^{3} f(t)\, dt$

$G(0) = G(-2) + \displaystyle\int_{-2}^{0} f(t)\, dt$

Therefore $G(3) - G(0) = \displaystyle\int_{-2}^{3} f(t) + \displaystyle\int_{-2}^{0} f(t)\, dt = \displaystyle\int_0^3 f(t)\, dt$

$G(3) = G(0) + \displaystyle\int_0^3 f(t)\, dt = 7 + 0 = 7$

Sample Examination II

1. $\int_0^2 (2x^3 + 3)\, dx = \left(\frac{2x^4}{4} + 3x \right) \Big|_0^2$

$\qquad\qquad\qquad\quad = \left(\frac{x^4}{2} + 3x \right) \Big|_0^2$

$\qquad\qquad\qquad\quad = (8 + 6) - (0 + 0) = 14$

The correct choice is (C).

2. The average value of f' is given by $\frac{1}{2-(-2)} \int_{-2}^2 f'(x)\, dx$. Using the Fundamental Theorem of

Calculus, $\frac{1}{2-(-2)} \int_{-2}^2 f'(x)\, dx = \frac{f(2) - f(-2)}{4} = \frac{e - a}{4}$

The correct choice is (B).

3. You need to find t such that $\frac{dy}{dx} = 4$.

$\dfrac{dy}{dx} = \dfrac{\frac{dy}{dt}}{\frac{dx}{dt}} = \dfrac{2t - 3}{4t + 1}$

Solving $\dfrac{2t - 3}{4t + 1} = 4$, you get $4(4t + 1) = 2t - 3$ or $16t + 4 = 2t - 3$, and $t = -\dfrac{1}{2}$.

The correct choice is (A).

4. $f(x) = (2 + 3x)^4$

$\quad f'(x) = 2(2 + 3x)^3 (3)$

$\quad f''(x) = 3 \cdot 4 (2 + 3x)^2 (3)^2$

$\quad f'''(x) = 2 \cdot 3 \cdot 4 (2 + 3x)(3)^3$

$\quad f^{(4)}(x) = 1 \cdot 2 \cdot 3 \cdot 4 \cdot (3)^4 = 4!\, (3^4)$

If $y = (a + bx)^n$, then the nth derivative, $\dfrac{d^n y}{dx^n} = n!\, (b^n)$

The correct choice is (C).

5. $f(x) = x^4 - 8x^2$

$f'(x) = 4x^3 - 16x = 4x(x^2 - 4)$

$f'(x) = 0$ when $x = 0, -2,$ and 2.

	$x < -2$	$-2 < x < 0$	$0 < x < 2$	$x > 2$
f'	$-$	$+$	$-$	$+$

$f(x)$ has a relative minimum where $f'(x)$ switches sign from $-$ to $+$.

Therefore, $f(x)$ has a relative minimum at $x = 2$ and $x = -2$ only.

The correct choice is (D).

6. $\displaystyle\int \sqrt{x}\,(x+2)\,dx = \int (x^{\frac{3}{2}} + 2x^{\frac{1}{2}})\,dx$

$\qquad\qquad\qquad\quad = \dfrac{2}{5}x^{\frac{5}{2}} + 2\cdot\dfrac{2}{3}x^{\frac{3}{2}} + C$

$\qquad\qquad\qquad\quad = \dfrac{2}{5}x^{\frac{5}{2}} + \dfrac{4}{3}x^{\frac{3}{2}} + C$

The correct choice is (D).

7. A rational function may have only one horizontal asymptote. If it does, then as $x \to \pm\infty$ the function must approach this value. This is expressed by the limit in (B). For example, the graph of $f(x) = \dfrac{7x^2}{x^2 + 3}$ is such a function. Its graph and asymptote are shown below:

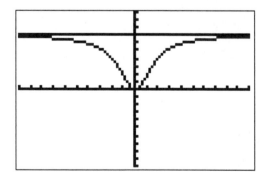

The correct choice is (B).

8. If the function $y = x^4 + bx^2 + 8x + 1$ has a horizontal tangent for some value of x, then the slope of y, y', is equal to 0 for that value of x.

$$y' = 4x^3 + 2bx + 8 \Rightarrow 4x^3 + 2bx + 8 = 0$$

If the function y has a point of inflection for some value of x, then $y'' = 0$ for that value of x.

$$y'' = 12x^2 + 2b \Rightarrow 12x^2 + 2b = 0 \Rightarrow 2b = -12x^2$$

Since the function has a horizontal tangent and a point of inflection for the same value of x,

$$4x^3 + 2bx + 8 = 12x^2 + 2b$$

Substituting $-12x^2$ for $2b$,

$$4x^3 + (-12x^2)(x) + 8 = 12x^2 + (-12x^2)$$
$$4x^3 - 12x^3 + 8 = 0$$
$$-8x^3 = -8$$
$$x^3 = 1$$
$$x = 1$$

Since $x = 1$ and $2b = -12x^2$, then $2b = -12$, $b = -6$.

Therefore the value of b is –6.

The correct choice is (A).

9. Euler's method iterates the formulas

$$x_{n+1} = x_n + \Delta x \text{ and } y_{n+1} = y_n + \frac{dy}{dx}\Delta x$$

$$\text{Thus,} y(5.5) \approx 1 + (1 - 5)(0.5) = -1$$

$$y(6) \approx -1 + (-1 - 5.5)(0.5) = -4.25$$

The correct choice is (A).

10. By observing the given graph of $f(x)$ and $T(x)$, the interval of convergence clearly includes 0.5 but not 1.5, so I is true and II is false. All Taylor polynomials are equal to the function at the center of the interval of convergence, so III is true.

The correct choice is (D).

11. Substitutions give the indeterminate form $\frac{0}{0}$, so using L'Hôpital's Rule:

$$\lim_{x \to 2} \frac{x^2 - 4}{\displaystyle\int_2^x \cos{(\pi t)}\, dt} = \lim_{x \to 2} \frac{2x}{\cos{(\pi x)}} = \frac{2\,(2)}{\cos{(2\pi)}} = \frac{4}{1} = 4$$

Note that by the Fundamental Theorem of Calculus: $\dfrac{d}{dx}\displaystyle\int_2^x \cos{(\pi t)}\, dt = \cos{(\pi x)}$.

The correct choice is (D).

12. I. $\displaystyle\int_0^{\infty} e^{-x}\, dx = \lim_{t \to +\infty}\int_0^t e^{-x}\, dx = \lim_{t \to +\infty}\left(-e^{-x}\right)\Big|_0^t = \lim_{t \to +\infty}\left(-e^{-t}\right) + 1 = 1$

   II. $\displaystyle\int_0^1 \frac{1}{x^2}\, dx = \lim_{t \to 0^+}\int_t^1 \frac{1}{x^2}\, dx = \lim_{t \to 0^+}\left(-\frac{1}{x}\right)\Big|_t^1 = -1 - \lim_{t \to 0^+}\left(-\frac{1}{t}\right) \to +\infty;$

   The integral diverges.

   III. $\displaystyle\int_0^1 \frac{1}{\sqrt{x}}\, dx = \lim_{t \to 0^+}\int_t^1 \frac{1}{\sqrt{x}}\, dx = \lim_{t \to 0^+} 2\sqrt{x}\,\Big|_t^1 = 2 - \lim_{t \to 0^+} 2\sqrt{t} = 2$

Therefore, only I and III converge.

The correct choice is (E).

13. $x + y = xy$

By implicit differentiation, $1 + \dfrac{dy}{dx} = x\left(\dfrac{dy}{dx}\right) + y$

$$\text{Solving for } \frac{dy}{dx}, \qquad 1 - y = x\left(\frac{dy}{dx}\right) - \frac{dy}{dx}$$
$$1 - y = \frac{dy}{dx}\,(x - 1)$$
$$\frac{dy}{dx} = \frac{1 - y}{x - 1}$$

The correct choice is (C).

14. Let $u = g(x)$ and $du = g'(x)\, dx$, then $\displaystyle\int_2^4 f'(g(x))\, g'(x)\, dx = \int f'(u)\, du = f(u) + C$

Thus $f(g(x))\Big|_2^4 = f(g(4)) - f(g(2))$

The correct choice is (D).

15. Integrate the velocity function $v(t)$ to find the position function $y(t)$:

$$y(t) = \int v(t)\, dt = \int (8 - 2t)\, dt = 8t - t^2 + C$$

The velocity is positive when $t < 4$ since during this time the particle is moving upwards.

At $t = 4$, $v(4) = 0$ and the particle stops moving up when it reaches the origin and begins moving down, therefore $y(4) = 0$.

$$y(4) = 8(4) - 4^2 + C = 0 \Rightarrow C = -16$$

Therefore, the position function is $y(t) = -t^2 + 8t - 16$.

The correct choice is (A).

16. Using the substitution $u = \sqrt{x - 1}$

$$u^2 = x - 1 \Rightarrow x = u^2 + 1 \text{ and } dx = 2u\, du$$

Use $u = \sqrt{x - 1}$ to change the limits of integration,

thus when $x = 2$, $u = 1$, and when $x = 5$, $u = 2$.

So $\displaystyle \int_2^5 \frac{\sqrt{x - 1}}{x}\, dx = \int_1^2 \frac{u}{u^2 + 1} \cdot 2u\, du = \int_1^2 \frac{2u^2}{u^2 + 1}\, du$

The correct choice is (E).

17. $e^x = 1 + x + \dfrac{x^2}{2!} + \cdots$

$e^{-x} = 1 - x + \dfrac{x^2}{2!} - \cdots$

$xe^{-x} = x\left(1 - x + \dfrac{x^2}{2!} - \cdots\right) = x - x^2 + \dfrac{x^3}{2!} - \cdots$

The correct choice is (B).

18. $\int_0^2 (2x^3 - kx^2 + 2k)\, dx = \frac{2x^4}{4} - \frac{kx^3}{3} + 2kx \Big|_0^2 = \left(\frac{32}{4} - \frac{8k}{3} + 4k\right) - (0) = 8 - \frac{8k}{3} + 4k$

Since $\int_0^2 (2x^3 - kx^2 + 2k)\, dx = 12$

Therefore, $8 - \frac{8k}{3} + 4k = 12$

$$24 + 4k = 36$$
$$k = 3$$

The correct choice is (E).

19. $\sum_{n=1}^{\infty} \left(\frac{1}{2}\right)^{2n}$ is a geometric series, i.e. $\left(\frac{1}{2}\right)^2 + \left(\frac{1}{2}\right)^4 + \left(\frac{1}{2}\right)^6 + \cdots$

where the first term is $\left(\frac{1}{2}\right)^2$ or $\frac{1}{4}$ and the ratio is also $\left(\frac{1}{2}\right)^2$ or $\frac{1}{4}$.

The sum of a geometric series is $\frac{a}{1-r}$ where a is the first term in the series, and r is the ratio.

Therefore, $\sum_{n=1}^{\infty} \left(\frac{1}{2}\right)^{2n}$ is the sum of the geometric series, which is $\frac{\frac{1}{4}}{1 - \frac{1}{4}} = \frac{1}{3}$

The correct choice is (A).

20. In general, the derivative of $y = \ln u$ is $y' = \frac{du}{u}$.

The derivative of $y = \ln\sqrt{1 - x^2}$ is

$$y' = \frac{1}{\sqrt{1 - x^2}} \cdot d\left(\sqrt{1 - x^2}\right) = \frac{1}{\sqrt{1 - x^2}} \cdot \frac{1}{2} \cdot \frac{1}{\sqrt{1 - x^2}} \cdot -2x = \frac{-x}{1 - x^2}$$

By multiplying both numerator and denominator by -1, the expression $\frac{-x}{1 - x^2} = \frac{x}{x^2 - 1}$. The

derivative of $y = \ln\sqrt{x^2 - 1}$ is $y' = \frac{x}{x^2 - 1}$.

The correct choice is (B).

21. A function whose derivative is a constant multiple of itself is equivalent to the expression $y' = ky$ (where k is the constant). The general solution of $y' = ky$ is $y = Ce^{kt}$. Therefore, the function must be exponential.

The correct choice is (C).

22. The graph is concave downward when $y'' < 0$.

$y = x^3 - 6x^2$, $y' = 3x^2 - 12x$, $y'' = 6x - 12$, $y'' < 0$ when $6x - 12 < 0$ or $x < 2$

Hence, $y = x^3 - 6x^2$ is concave downward when $x < 2$.

The correct choice is (C).

23. By the quotient rule the integrand $\dfrac{f(x)\,g'(x) - g(x)\,f'(x)}{(f(x))^2} = \dfrac{d}{dx}\left(\dfrac{g(x)}{f(x)}\right)$. Therefore

$$\int_2^8 \frac{f(x)\,g'(x) - g(x)\,f'(x)}{(f(x))^2}\,dx = \int_2^8 \frac{d}{dx}\left(\frac{g(x)}{f(x)}\right)dx$$

$$= \left.\frac{g(x)}{f(x)}\right|_2^8$$

$$= \frac{g(8)}{f(8)} - \frac{g(2)}{f(2)}$$

$$= \frac{-6}{3} - \frac{2}{1} = -4$$

The correct choice is (E).

24. I. The series may be written as $1 + \dfrac{1}{2^{3/2}} + \dfrac{1}{3^{3/2}} + \cdots + \dfrac{1}{n^{3/2}} + \cdots$

   This series is convergent because it is a p-series with $p > 1$.

   II. Compare the given series with the convergent geometric series $(r = 1/2)$,

   $\dfrac{1}{2} + \dfrac{1}{4} + \dfrac{1}{8} + \cdots + \dfrac{1}{2n} + \cdots$.

   Each term of the given series is equal to or less than the corresponding term of the convergent geometric series. Therefore the given series is convergent.

   (The p-series, $\dfrac{1}{n^2}$ could also be used in the Comparison Test.)

   III. Compare the given series with $1 + \dfrac{1}{2} + \dfrac{1}{3} + \cdots + \dfrac{1}{n} + \cdots$ Each term of the given series is equal to or greater than the corresponding term of the divergent harmonic series, therefore this series also diverges.

The correct choice is (C).

25. $f'(x) = \displaystyle\sum_{n=0}^{\infty} \frac{(-1)^{n+1}(2n+1)\,x^{2n}}{(2n+1)!} = \sum_{n=0}^{\infty} \frac{(-1)^{n+1}x^{2n}}{(2n)!}$, since $(2n+1)! = (2n+1)(2n)!$

The correct choice is (B).

26. $y = \sqrt[3]{x^2-1} = (x^2-1)^{\frac{1}{3}}$

$y' = \dfrac{1}{3}(x^2-1)^{-\frac{2}{3}}(2x)$

$y'(3) = \left(\dfrac{1}{3}\right)\left(\dfrac{1}{4}\right)(6) = \dfrac{1}{2}$

Thus the <u>normal</u> line to the curve will have a slope of -2.

The equation of the normal line to the curve $y = \sqrt[3]{x^2-1}$ at $(3,2)$

is $y - 2 = -2(x-3)$ or $y + 2x = 8$

The correct choice is (D).

27. Since the velocity, $s'(t)$, and the acceleration, $s''(t)$, are both positive, the graph of $s(t)$ will be increasing $(s'(t) > 0)$ and concave upwards $(s''(t) > 0)$. Only the graph of (C) satisfies both of these conditions.

The correct choice is (C).

28. The maximum value of a function occurs where its first derivative changes from positive to negative <u>or</u> at an endpoint of the domain of the function. By the Fundamental Theorem of Calculus $g'(x) = f(x)$. From the graph $f(x)$ changes from positive to negative at $x = 3$. Thus the maximum value of $g(x)$ occurs at $x = 3$.

<u>Alternate Method</u>: Evaluate $g(x)$ as area of the regions of $f(t)$ to the x-axis. Those regions below the x-axis are counted as negative values.

Then $g(-3) = \displaystyle\int_0^{-3} f(t)\,dt = -8,$

$\qquad g(3) = \displaystyle\int_0^3 f(t)\,dt = 3.5,$

and $\quad g(5) = \displaystyle\int_0^5 f(t)\,dt = 1.5$

Thus the maximum value of $g(x)$ occurs at $x = 3$.

The correct choice is (D).

29. Differentiate the position equations to find the velocity vector:

$$v(t) = \langle -9 \sin t, 4 \cos t \rangle.$$

Then differentiate the velocity vector to find the acceleration vector:

$$a(t) = \langle -9 \cos t, -4 \sin t \rangle$$

Substitute the value $t = 3$, $a(3) = \langle -9 \cos (3), -4 \sin (3) \rangle = \langle 8.190, -0.564 \rangle.$

The correct choice is (C).

30. The graph of $f'(x)$ is always increasing, thus $f''(2) > 0$.

Since the derivative changes from negative to positive at $x = 2$, $f(2)$ is a minimum value. Since $f(1) = -2$ and $f(2) < f(1)$, then $f(2) < -2$.

From the graph, $f'(2) = 0$.

So arranging the numbers in order $f(2) < f'(2) < f''(2)$.

The correct choice is (A).

31. Since $f'(x)$ is always increasing and changes sign from negative to positive at $x = 0$, the First Derivative Test tells us that $x = 0$ is a relative minimum. (A) must be true.

Since $f'(x)$ is always increasing, $f(x)$ is always concave upwards and has no point of inflection. Hence, (B) and (C) are not true.

Since $f(0)$ is not given it cannot be determined if $f(x)$ passes through the origin — therefore, (D) need not be true.

Since the derivative $f'(x)$ is <u>not</u> even $((f'(10) \neq f'(-10)))$, then $f(x)$ cannot be odd. The derivative of an odd function must be even. So (E) is not true.

The correct choice is (A).

32. Using a calculator, the amount is found by integrating $M(t)$: $\int_0^3 (4 - (\sin t)^3) \, dt \approx 10.667.$

The correct choice is (B).

33. To the right of the origin the function is decreasing; therefore, its derivative $f'(x) < 0$, and I is true. As $x \to +\infty$ and $x \to -\infty$, the graph flattens out so its slope $f'(x)$ is approaching zero; II is true and III is false.

    The correct choice is (D).

34. Integrate the velocity vector to obtain the position vector: $\langle 2t^2 + C, -t^2 + K \rangle$. Then graph this on the graphing calculator. Regardless of the constants of integration (try any values you want) the graph will be a ray. Note that at $t = 0$ the ray's vertex is (C, K).

    Alternatively, $\dfrac{dy}{dx} = \dfrac{\frac{dy}{dt}}{\frac{dx}{dt}} = \dfrac{-2t}{4t} = -\dfrac{1}{2}$ is constant so the curve must be either a line or a ray. Since t is limited to non-negative numbers, the curve is a ray.

    The correct choice is (E).

35. The acceleration becomes negative when the velocity stops increasing and starts decreasing (even though the velocity may still be positive). This corresponds to a point of inflection on the graph, where the concavity changes from up (on the left) to down (on the right). Point C appears to be the closest point where there is a point of inflection.

    The correct choice is (C).

36. Graph $f'(x)$ in the window $[0,5]$ by $[-50,5]$. The derivative changes from negative (indicating a decreasing function) to positive (increasing function) at $x = 0.618$, thus indicating a relative minimum. Now check how the endpoint minimum at $x = 5$ compares to this value. This may be done by evaluating $\displaystyle\int_{0.618}^{5} [e^x(-x^3 + 3x) - 3]\, dx$ on a graphing calculator or by looking at the area above and below the x-axis between $x = 0.618$ and $x = 5$. Since there is more area below or by evaluating the definite integral, the conclusion is that $F(5) - F(0.618) < 0$ and that $F(5) < F(0.618)$.

    The correct choice is (E).

37. The derivative may be approximated by calculating the slope from any two points near $x = 4$, for example,

$$f'(4) \approx \frac{1.16016 - 1.15782}{4.00100 - 4.00000} = \frac{0.00234}{0.001} = 2.340$$

Using other points gives comparable results.

The correct choice is (D).

38. The length of the path $= \int_0^{2\pi} \sqrt{\left(\frac{dx}{dt}\right)^2 + \left(\frac{dy}{dt}\right)^2} \, dt$

$\frac{dx}{dt} = \cos t$ and $\frac{dy}{dt} = -2\sin(2t)$

Length $= \int_0^{2\pi} \sqrt{\cos^2 t + 4\sin^2(2t)} \, dt$

Use a calculator to find the length to be 9.294.

The correct choice is (B).

39. Graph the two functions $y = \cos 2x$ and $y = \sin 3x$ in the interval $0 \leq x \leq 5$. Choices (A) through (D) represent the areas of the regions enclosed by the graphs from left to right. It is easy to see that choice (D) is the largest. Choice (E) has the wrong function "on top" and therefore represents a negative number.

The correct choice is (D).

40. First graph the function $f(x)$ in the window $[0, 1.4]$ by $[-10, 10]$. Identify the zero at $x = 1.042$ (which is $\frac{\ln \pi}{\ln 3}$). Then use the derivative function to find the derivative; and, the derivative at the point ($x = 1.042$) is 3.451. Do not be misled by the vertical asymptote near $x = 0.411$.

The correct choice is (C).

41. The interval $[0,6]$ is divided into three subintervals $[0,2]$, $[2,4]$, and $[4,6]$. The integral is approximated using the function's value at the midpoint of each interval multiplied by the width of the interval:

$$(0.25)(2) + (0.68)(2) + (0.95)(2) = 3.76$$

For additional practice: (A) is the left Riemann sum, (E) is the right Riemann sum and (B) is the trapezoidal approximation, all with three subintervals.

The correct choice is (D).

42. By the Fundamental Theorem of Calculus (version II),

$$\frac{d}{dx} \int_x^{x^3} \sin(t^2)\, dt = 3x^2 \sin((x^3)^2) - \sin(x^2)$$

$$= 3x^2 \sin(x^6) - \sin(x^2)$$

The correct choice is (C).

43. The amount of water, A, after x hours, is given by the accumulation function

$$A(x) = \int_0^x 300\sqrt{t}\, dt$$

After 4 hours, there are $\int_0^4 300\sqrt{t}\, dt$ number of gallons.

$$\int_0^4 300\sqrt{t}\, dt = 300 \int_0^4 \sqrt{t}\, dt = 300 \cdot \frac{2}{3} \cdot t^{\frac{3}{2}} \Big|_0^4 = 200\left(4^{\frac{3}{2}} - 0\right) = 1600 \text{ gallons}$$

The correct choice is (D).

44. Solve $2\sin x = x$ to obtain $x = 1.89549$. Let $A = 1.89549$.

Using a graphing calculator, evaluate $V = \pi \int_0^A [(2\sin x)^2 - x^2]\, dx = 6.678$.

The correct choice is (D).

45.

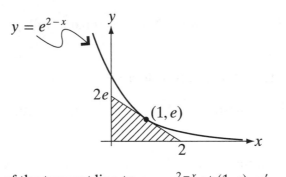

First, find the equation of the tangent line to $y = e^{2-x}$ at $(1, e)$. $y' = -e^{2-x}$; therefore, at $x = 1$, $y'(1) = -e$. The equation of the tangent line is $y - e = -e(x - 1)$ or $y = -ex + 2e$. To find the area of the triangle (the shaded region) you need the base and height of the triangle; these are the x-intercept and the y-intercept of the tangent line. The x-intercept is $x = 2$ and the y-intercept is $y = 2e$.

Therefore, the area of the triangle $= \frac{1}{2}$ (base) (height) $= \frac{1}{2}(2)(2e) = 2e$.

The correct choice is (A).

1a. The distance is given by $\int_0^5 (-5t^2 + 20t + 25)\, dt = 166.666$ miles.

1b. The acceleration is the derivative of the velocity:

$$\frac{d}{dt} v(t)\Big|_{t=3} = -10t + 20\Big|_{t=3} = -10(3) + 20 = -10 \text{ miles/hour/hour}$$

1c. Any one of the following approaches below is reasonable. Be sure to show your work so that the method you use is clear.

Using a left Riemann sum with $n = 5$, the distance traveled in the second part of the trip is $(1 \text{ hour})(0 + (-25) + (-20) + (-50) + k) \text{ miles/hour} = (-95 + k) \text{ miles}$. So $-95 + k \approx -167$ and $k \approx -72$ miles per hour.

Using a right Riemann sum with $n = 5$, gives $(1 \text{ hour})((-25) + (-20) + (-50) + k + (-20)) \text{ miles per hour} = (-115 + k) \text{ miles}$. So $-115 + k \approx -167$ and $k \approx -52$ miles per hour.

2a. Graph the function $T(x)$ in a suitable window such as $[0,24]$ by $[50,100]$. Using a graphing calculator, find the value of x which makes $y = 70$. Rounded to the nearest half-hour, $x \approx 8.5$ or at 8:30 am.

2b. Same idea as in part (a). Find the value of x which makes $y = 77$. Rounded to the nearest half-hour, $x \approx 10.5$ or 10:30 am.

2c. Use the cost equation with $x = 8.5$ and $T_0 = 70$:

$$C(x) = 0.16 \int_{8.5}^{18} (T(x) - 70) \, dx = 0.16 \int_{8.5}^{18} \left[73 - 14 \cos\left(\frac{\pi(x - 3.4)}{12} \right) - 70 \right] dx = \$18.26$$

Use a graphing calculator to evaluate the definite integral.

2d. In a like manner, use $T_0 = 77$ and $x = 10.5$, to first find the cost of cooling at 77°F:

$$C(x) = 0.16 \int_{10.5}^{18} \left[73 - 14 \cos\left(\frac{\pi(x - 3.4)}{12} \right) - 77 \right] dx = \$8.79$$

The savings is $\$18.26 - \$8.79 = \$9.47$ per day.

REMINDER: Numeric and algebraic answers need not be simplified, and if simplified incorrectly credit will be lost.

3a. At $(5,6)$ the slope of the tangent line is $h'(5) = 0.7$. The tangent line is

$$y(x) = 0.7(x-5) + 6$$
$$y(4) = 0.7(-1) + 6$$
$$= 5.3 \text{ inches}$$

Since $h''(t) > 0$, the function is concave up, so the tangent line lies below the curve and gives an approximation that is less than the true value.

3b. $V = (10)(20)h$

$$\frac{dV}{dt} = 200 \frac{dh}{dt}$$

$$\left.\frac{dV}{dt}\right|_{t=2} = 200(0.5) = 100 \text{ cubic inches per minute.}$$

3c. $h \approx 0.4(2) + 0.5(3) + 0.7(4) + 1.0(4) + 1.1(2)$

≈ 11.3 inches

$\int_0^{15} h'(t)\, dt$ is the change in the depth of water in the tank, in inches, from $t = 0$ to $t = 15$ minutes.

(Note: $\int_0^{15} h'(t)\, dt$ does *not* represent the depth of water in the tank at $t = 15$ minutes because the depth of water at $t = 0$ minutes is unknown).

3d. The approximation in part (c) is less than the true amount. Since $h''(t) > 0, h'(t)$ is increasing, and the left side values used in the approximation are all less than the true values.
(Note: For a function which is increasing and concave up, the rectangles in a left Riemann sum all lie below the curve.)

4a. $y = \dfrac{x}{x+C}, x \neq C, C \neq 0$ and from the given differential equation: $y' = \dfrac{y - y^2}{x}$

$$y' = \frac{(x+C)(1) - x(1)}{(x+C)^2}$$

$$y' = \frac{C}{(x+C)^2}$$

$$= \frac{\frac{x}{x+C} - \frac{x^2}{(x+C)^2}}{x}$$

$$= \frac{x(x+C) - x^2}{x(x+C)^2}$$

$$= \frac{C}{(x+C)^2}$$

By substituting $(0,0)$ into $y = \dfrac{x}{x+C}: 0 = \dfrac{0}{0+C}$ is true, so regardless of the value of C, $(0,0)$ is on the solution curve.

4b. Substituting $(1,2)$ into $y = \dfrac{x}{x+C}$ gives $2 = \dfrac{1}{1+C}$ and $C = -\dfrac{1}{2}$. The particular solution is

$y = \dfrac{x}{x - \frac{1}{2}}$, and the slope at $(0,0)$ is $y'(0) = \dfrac{-\frac{1}{2}}{\left(0 - \frac{1}{2}\right)^2} = -2$

4c. The asymptotes are $x = \dfrac{1}{2}$ and $y = 1$.

4d.

5a. $v(t) = \langle x'(t), y'(t) \rangle = \langle 2, -2t \rangle$

5b. $\dfrac{dy}{dx} = \dfrac{dy/dt}{dx/dt} = \dfrac{-2t}{2} = -t$ is the slope at any point. Then the equation is

$y - (36 - t^2) = (-t)(x - 2t)$ or $y = -t(x - 2t) + (36 - t^2)$

5c. The line from part (b) intersects the x-axis when $y = 0$. Substitute this value into the equation of the tangent line and solve for x:

$$\begin{aligned} 0 &= -tx + 2t^2 + 36 - t^2 \\ tx &= t^2 + 36 \\ x &= \frac{t^2 + 36}{t} = t + 36t^{-1} \end{aligned}$$

5d. The speed of the laser light moving along the x-axis is obtained by taking $\left|\dfrac{dx}{dt}\right|$ obtained from part (c).

$\left|\dfrac{dx}{dt}\right| = \left|1 - \dfrac{36}{t^2}\right| = \dfrac{36 - t^2}{t^2}$.

At $t = 3$, $\left|\dfrac{dx}{dt}\right| = \left|1 - \dfrac{36}{3^2}\right| = |-3| = 3$.

6a. At any point the slope is $\dfrac{df}{dx} = -2x$. The tangent line's equation is

$$\begin{aligned}
y - (4 - a^2) &= -2a(x - a) \\
y &= -2ax + 2a^2 + 4 - a^2 \\
y &= -2ax + a^2 + 4
\end{aligned}$$

6b. Using the answer from (a) the area is the integral of the upper curve (the line) minus the lower:

$$A = \int_0^a \left((-2ax + a^2 + 4) - (4 - x^2) \right) dx$$

6c. Evaluating the integral in (b):

$$A = \int_0^a \left((-2ax + a^2 + 4) - (4 - x^2) \right) dx$$

$$A = -ax^2 + a^2 x + 4x - 4x + \frac{1}{3}x^3 \Big|_0^a$$

$$A = -a^3 + a^3 + 4a - 4a + \frac{1}{3}a^3$$

$$A = \frac{1}{3}a^3$$

6d. $A(x) = \dfrac{1}{3}x^3$

The average value of the area of region R on the interval $0 \le x \le a$ is

$$\frac{1}{a - 0} \int_0^a \frac{1}{3}x^3\,dx = \frac{1}{12a}x^4 \Big|_0^a = \frac{1}{12a}(a^4 - 0) = \frac{a^3}{12}$$

Sample Examination III

1. The area of the region is represented by $\int_1^3 (3x^2 + 2x)\, dx$.

$$\int_1^3 (3x^2 + 2x)\, dx = x^3 + x^2 \big|_1^3 = (27 + 9) - (1 + 1) = 36 - 2 = 34$$

The correct choice is (B).

2. The given limit is in indeterminate form, $\frac{0}{0}$, so L'Hôpital's Rule applies.

$$\lim_{x \to 0} \frac{x}{e^{2x} - 1} = \lim_{x \to 0} \frac{1}{2e^{2x}} = \frac{1}{2e^0} = \frac{1}{2}$$

The correct choice is (B).

3. The units of a definite integral or a Riemann sum are the units of the dependent variable (the integrand) multiplied by the units of the independent variable, ds, in this case.

$$(\text{gallons/mile}) \times (\text{miles/hour}) = \text{gallons/hour}$$

The correct choice is (C).

4. The velocity vector is found by finding $\frac{dx}{dt}$ and $\frac{dy}{dt}$.

$$\frac{dx}{dt} = (e^t)(\cos t) + (\sin t)(e^t) \text{ and } \frac{dy}{dt} = (e^t)(-\sin t) + (\cos t)(e^t)$$

At $t = \pi$, $\dfrac{dx}{dt} = (e^\pi)(\cos \pi) + (\sin \pi)(e^\pi) = -e^\pi + 0 = -e^\pi$

At $t = \pi$, $\dfrac{dy}{dt} = (e^\pi)(0) + (-1)(e^\pi) = -e^\pi$

Therefore, the velocity vector $v(t) = \langle -e^\pi, -e^\pi \rangle$.

The correct choice is (E).

5. Since the velocity is positive over the given interval, the distance the particle travels from $t = 1$ to $t = 3$ can be represented by

$$\int_1^3 v(t)\, dt = \int_1^3 t^2\, dt = \frac{t^3}{3} \Big|_1^3 = 9 - \frac{1}{3} = \frac{26}{3}$$

The correct choice is (B).

6. By the Fundamental Theorem of Calculus: $\int_1^4 f'(x)\, dx = f(4) - f(1) = 2 - (-5) = 7$

   The correct choice is (D).

7. The choices give three different initial conditions for the unknown differential equation. For each trace along the slope field and see where the curves go. Starting above $(0,2)$ the path moves down to the right leveling off near $y = 2$. So I is true. Starting between 0 and 2 on the y-axis and moving right the curve rises towards $y = 2$, so II is true. Starting on the y-axis below the origin and tracing to the right leads away from the x-axis, so III is false.

   The correct choice is (D).

8. Let $u = x^2 + 1$

   $$du = 2x\, dx \Rightarrow x\, dx = \frac{1}{2} du$$

   $$\int \frac{x}{x^2 + 1}\, dx = \frac{1}{2} \int \frac{du}{u} = \frac{1}{2} \ln|u| = \frac{1}{2} \ln(x^2 + 1)\Big|_1^3 = \frac{1}{2}(\ln 10 - \ln 2)$$

   Since $\ln 10 - \ln 2 = \ln\left(\frac{10}{2}\right) = \ln 5$, therefore, $\int_1^3 \frac{x}{x^2 + 1}\, dx = \frac{1}{2} \ln 5$.

   The correct choice is (D).

9. Using the properties of logarithms and the Chain Rule:

   $$\frac{d}{dx} \ln\left(\frac{1}{x^2 - 1}\right) = \frac{d}{dx}\left(\ln 1 - \ln(x^2 - 1)\right)$$

   $$= \frac{d}{dx}\left(-\ln(x^2 - 1)\right)$$

   $$= -\frac{2x}{x^2 - 1} = \frac{2x}{1 - x^2}$$

   The correct choice is (B).

10. $\int_{1}^{\infty} x^{-\frac{5}{4}} dx$ is an improper integral.

$$\int_{1}^{\infty} x^{-\frac{5}{4}} dx = \lim_{t \to \infty} \int_{1}^{t} x^{-\frac{5}{4}} dx = \lim_{t \to \infty} \left(\frac{x^{-\frac{1}{4}}}{-\frac{1}{4}} \right) \Big|_{1}^{t}$$

$$= \lim_{t \to \infty} \left(\frac{t^{-\frac{1}{4}}}{-\frac{1}{4}} - \frac{1^{-\frac{1}{4}}}{-\frac{1}{4}} \right)$$

$$= 0 - \frac{1}{-\frac{1}{4}} = 4$$

The correct choice is (A).

11. Since the function is concave down, the first derivative must always decrease.

In table (A): $\dfrac{f(5) - f(4)}{5 - 4} = -4$ so by the Mean Value Theorem there must be a value, c_1, between $x = 4$ and $x = 5$ where $f'(c_1) = -4$. Also $\dfrac{f(6) - f(5)}{6 - 5} = -4$ so there must be a value c_2, between $x = 5$ and $x = 6$ where $f'(c_2) = -4$. Thus, the derivative goes from -4 to -3 (given) to -4. Since the derivative is not decreasing throughout the interval, (A) cannot be the correct table.

Doing a similar analysis on the other tables, the derivatives, in the same order, are

(B) $-2, -3, -4$

(C) $-2, -3, -1$

(D) $-2, -3, -2$

(E) $-5, -3, -1$

This eliminates (C), (D) and (E). The remaining choice (B) is the only one that *could* be the table of values for f.

Although what happens between the values given in the table is unknown, the slopes in (B) clearly indicate a decrease from -2 to -4 and therefore (B) is the <u>best</u> possible choice. In all of the other four choices it is clear that the slopes do not continually decrease.

The correct choice is (B).

12. $y' = \dfrac{k(k+x)-(kx+8)}{(k+x)^2}$, or

$y' = \dfrac{k^2-8}{(k+x)^2}$

Since $y = x + 4$ is the line tangent to y at $x = -2$, its slope is $y' = 1$.

By substituting $x = -2$ and $y'(-2) = 1$,

$$1 = \frac{k^2-8}{(k-2)^2} \Rightarrow k^2 - 8 = (k-2)^2 \text{ or } k^2 - 8 = k^2 - 4k + 4, \text{ and } k = 3.$$

The correct choice is (D).

13. Notice that the subintervals are not of equal width. Multiplying the function on the right side of each subinterval times the width of that subinterval gives

$$7(1) + 11(3) + 12(2) + 8(1) = 72$$

The correct choice is (E).

14. The series is a geometric series with a ratio of $\left(-\dfrac{\pi}{3}\right)$.

Since the absolute value of the ratio is larger than 1, the series will not converge.

The correct choice is (E).

15. By the First Derivative Test, for a function to have a minimum value the derivative must change sign from negative to positive, this is choice (D). Choices (A) and (C) do not have enough information to justify a maximum or minimum. Choice (B) is wrong because for a maximum, the derivative must change from positive to negative. For choice (E) to be a correct justification by the Second Derivative Test, you also need to know that $f'(3) = 0$.

The correct choice is (D).

16. At a point of inflection where the graph changes from concave up to concave down the *second* derivative ($g'' = f'$) changes from positive to negative. This occurs where the *first* derivative ($g' = f$) changes from increasing to decreasing. The value is $x = 5$. Since f has a relative maximum at $x = 5$, f' changes from positive to negative and thus g changes from concave up to concave down.

The correct choice is (D).

17. First, observe that $f(x)$ is increasing where $f'(x) \geq 0$ and is decreasing where $f'(x) < 0$. In the given graph of $f'(x)$, $f'(x) \geq 0$ for $x \geq -2$; $f'(x) < 0$ for $x < -2$. Therefore, $f(x)$ is decreasing for $x < -2$ and increasing for all $x \geq -2$. From the given choices, the only graph that is decreasing to the left of $x = -2$ and increasing to the right of $x = -2$ is (A).

    The correct choice is (A).

18. $\dfrac{dy}{dt} = \dfrac{1}{2}(2t + 5)^{-\frac{1}{2}} \cdot 2 = (2t + 5)^{-\frac{1}{2}}$ and $\dfrac{dx}{dt} = 1 - 2t$

    $\dfrac{dy}{dx} = \dfrac{dy/dt}{dx/dt} = \dfrac{(2t + 5)^{-\frac{1}{2}}}{1 - 2t}\bigg|_{t=2} = \dfrac{9^{-\frac{1}{2}}}{-3} = -\dfrac{1}{9}$

    The correct choice is (C).

19. The segment with endpoints $(-2, 26)$ and $(10, 2)$ has a slope of -2. Therefore, by the Mean Value Theorem, the line through $(4, 23)$ will be parallel to the segment and tangent to the graph; its equation is $y - 23 = -2(x - 4)$. Its y-intercept is 31.

    The correct choice is (C).

20. Since at each point (x, y) on a certain curve, the slope of the curve is $4xy$, $\dfrac{dy}{dx} = 4xy$.

    To find y (the equation of the curve), solve the following separable differential equation.

    $$\dfrac{dy}{dx} = 4xy \Rightarrow \dfrac{dy}{y} = 4x\, dx \Rightarrow \int \dfrac{dy}{y} = \int 4x\, dx$$

    $$\Rightarrow \ln|y| = 2x^2 + C_1$$

    $$\Rightarrow y = Ce^{2x^2}$$

    Since the curve contains the point $(0, 4)$, $4 = Ce^0 \Rightarrow C = 4$.

    Therefore, $y = 4e^{2x^2}$.

    The correct choice is (A).

21. Using integration by partial fractions,

$$\frac{1}{x^2 - 9} = \frac{1}{(x-3)(x+3)} = \frac{A}{x-3} + \frac{B}{x+3} \Rightarrow 1 = A(x+3) + B(x-3)$$

Let $x = 3$. Then $1 = 6A$ or $A = \frac{1}{6}$.

Let $x = -3$. Then $1 = -6B$ or $B = -\frac{1}{6}$.

So $\frac{1}{x^2 - 9} = \frac{1}{6}\left(\frac{1}{x-3}\right) - \frac{1}{6}\left(\frac{1}{x+3}\right)$

Therefore,

$$\int \frac{dx}{x^2 - 9} = \frac{1}{6}\int \frac{dx}{x-3} - \frac{1}{6}\int \frac{dx}{x+3} = \frac{1}{6}\ln|x-3| - \frac{1}{6}\ln|x+3| + C = \frac{1}{6}\ln\left|\frac{x-3}{x+3}\right| + C$$

The correct choice is (E).

22. The length of a continuous curve $y = f(x)$ from $x = a$ to $x = b$ is $L = \int_a^b \sqrt{1 + (y')^2}\, dx$.

In the given problem, $y = e^{(e^x)}$. So $y' = e^{(e^x)} \cdot e^x = e^{(x + e^x)}$ and $(y')^2 = e^{2(x + e^x)}$.

Therefore, $L = \int_0^1 \sqrt{1 + e^{2(x + e^x)}}\, dx$.

The correct choice is (A).

23. The only properties which are true are I and III.

Property II is false because only a constant may be "taken out" of the integrand.

The correct choice is (C).

24. $f(x) = \sqrt{e^{2x} + 1} = (e^{2x} + 1)^{\frac{1}{2}}$ and $f'(x) = \frac{1}{2}(e^{2x} + 1)^{-\frac{1}{2}} \cdot 2e^{2x} = \frac{e^{2x}}{\sqrt{e^{2x} + 1}}$

Therefore, $f'(0) = \frac{e^0}{\sqrt{e^0 + 1}} = \frac{1}{\sqrt{2}} = \frac{\sqrt{2}}{2}$

The correct choice is (C).

25. The given limit is actually the derivative of $\tan 2x$.

    If $f(x) = \tan 2x$, $f'(x) = \lim_{h \to 0} \dfrac{f(x + h) - f(x)}{h}$

    $$= \lim_{h \to 0} \frac{\tan(2(x + h)) - \tan(2x)}{h}$$

    Consequently, the derivative of $\tan 2x$ is $2 \sec^2(2x)$.

    The correct choice is (D).

26. Since every cross section is a semicircle, Area $= \frac{1}{2}\pi(\text{radius})^2 = \frac{1}{2}\pi\left(\frac{1}{2}y\right)^2 = \frac{\pi}{8}y^2$.

    $\text{Volume} = \displaystyle\int_0^4 \frac{\pi}{8}\left(\frac{4 - x}{2}\right)^2 dx$ (since $x + 2y = 4 \Rightarrow 2y = 4 - x$ or $y = \frac{4-x}{2}$)

    $$= \frac{\pi}{32}\int_0^4 (4 - x)^2 dx$$

    $$= \frac{\pi}{32}\int_0^4 (16 - 8x + x^2)\, dx$$

    $$= \frac{\pi}{32}\left(16x - 4x^2 + \frac{x^3}{3}\right)\Bigg|_0^4$$

    $$= \frac{\pi}{32}\left[\left(64 - 64 + \frac{64}{3}\right) - (0)\right]$$

    $$= \frac{\pi}{32}\left(\frac{64}{3}\right) = \frac{2\pi}{3}$$

    The correct choice is (A).

27. I diverges because its partial sums alternate between 1 and 0.

    II diverges because $\lim_{n \to \infty} (-1)^{n+1} n \neq 0$. ($n$th-Term Test for Divergence)

    III diverges because $\lim_{n \to \infty} \left(\dfrac{1 + n}{n}\right)^n = \lim_{n \to \infty} \left(1 + \dfrac{1}{n}\right)^n = e$.

    <u>Note</u>: since the limit of the nth term is not zero, the series diverges.

    The correct choice is (A).

28. The general solution to the differential equation $y' = ky$ is $y = Ce^{kt}$ where C is the initial population. It is given that the initial population, $C = 1,500$ and this population is 6,000 after the first 2 days.

$$6,000 = 1,500e^{2k}$$
$$4 = e^{2k}$$
$$\ln 4 = 2k$$
$$k = \frac{\ln 4}{2} = \frac{2 \ln 2}{2} = \ln 2$$
$$\text{Thus } y = 1,500e^{t \ln 2}$$
$$\text{At the time } t = 3, y = 1,500e^{3 \ln 2}$$
$$y = 1,500e^{\ln(2^3)}$$
$$y = 1,500(2^3)$$
$$y = 1,500(8)$$

Therefore, the population will have increased by a factor of 8.

The correct choice is (D).

29. The average value of $f(x)$ is $\dfrac{1}{7-3}\displaystyle\int_3^7 f(x)\, dx = 12$

$$\Rightarrow \frac{1}{4}\int_3^7 f(x)\, dx = 12$$

$$\Rightarrow \int_3^7 f(x)\, dx = 48$$

The correct choice is (E).

30. Using the ratio test, $\displaystyle\lim_{n\to\infty}\left|\frac{a_{n+1}}{a_n}\right| = \lim_{n\to\infty}\frac{\dfrac{2^{n+1}}{(n+2)!}}{\dfrac{2^n}{(n+1)!}} = \lim_{n\to\infty}\frac{2}{n+2} < 1$

The correct choice is (A).

31. Find the second derivative of f:

    $$f'(x) = 2x - 5 \sin x$$

    $$f''(x) = 2 - 5 \cos x$$

    Points of inflection occur where the second derivative changes sign. In this case, the second derivative is zero and changes sign when $\cos x = \frac{2}{5}$. In any interval of length 2π this happens twice. Therefore, in the given interval $0 \le x \le 2\pi n$, it will occur $2n$ times.

    The correct choice is (E).

32. Since the function has a derivative for all $x \ne 2$, it is continuous everywhere else. The derivative is positive indicating that the function is always increasing. The function must increase to infinity on the left side of the asymptote and then increase *from* $-\infty$ on the right side. In other words, this is an odd vertical asymptote. Thus the limit in III should be $-\infty$ and the limit in I does not exist since the two one-sided limits are different. Only II is true. An example of such a function is $f(x) = \frac{-1}{x-2}$; its graph is shown below:

    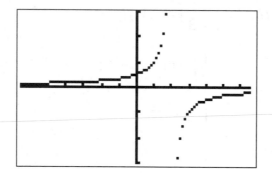

    The correct choice is (B).

33. By the Fundamental Theorem of Calculus, $f'(x) = \cos \sqrt{x}$.

    The frequently used series for $\cos x$ is

    $$\cos x = 1 - \frac{x^2}{2!} + \frac{x^4}{4!} - \frac{x^6}{6!} + \cdots$$

    $$f'(x) = \cos \sqrt{x} = 1 - \frac{(\sqrt{x})^2}{2!} + \frac{(\sqrt{x})^4}{4!} - \frac{(\sqrt{x})^6}{6!} + \cdots$$

    $$\text{or } \cos \sqrt{x} = 1 - \frac{x}{2!} + \frac{x^2}{4!} - \frac{x^3}{6!} + \cdots$$

    To find the Taylor series of f, integrate $f'(x)$.

$$\int f'(x)\, dx = f(x) = \int \left(1 - \frac{x}{2!} + \frac{x^2}{4!} - \frac{x^3}{6!} + \cdots\right) dx$$

$$= x - \frac{x^2}{4} + \frac{x^3}{72} - \frac{x^4}{2880} + \cdots$$

The correct choice is (D).

34. Speed $= \sqrt{(x'(t))^2 + (y'(t))^2}$

    $x(t) = 4\sin(\pi t)$ so $x'(t) = 4\pi \cos(\pi t)$

    $y(t) = (3t - 1)^2$ so $y'(t) = 2(3t - 1)(3)$ or $18t - 6$

    At $t = 2, x'(2) = 4\pi \cos(2\pi) = 4\pi$ and $y'(2) = 30$

    Therefore, the speed at $t = 2$ is $\sqrt{(4\pi)^2 + (30)^2} = 32.526$

    The correct choice is (D).

35. Graph the second derivative. The function $f(x)$ will have a point of inflection when the second derivative changes sign. The second derivative changes sign (crosses the x-axis) six times in this interval.

    The correct choice is (C).

36. The formula for arc length is $L = \int_a^b \sqrt{\left(\dfrac{dx}{dt}\right)^2 + \left(\dfrac{dy}{dt}\right)^2}\, dt.$

    The derivatives are: $dx/dt = t^3$ and $dy/dt = \cos t$.

    The length is $\int_0^{2\pi} \sqrt{(t^3)^2 + (\cos t)^2}\, dt = \int_0^{2\pi} \sqrt{t^6 + \cos^2 t}\, dt.$

    The correct choice is (D).

37. To approximate $f(2.2)$ center the Taylor polynomial about $x = 2$.

$$f(x) \approx f(2) + f'(2)(x - 2) + \frac{f''(2)}{2!}(x - 2)^2 + \frac{f'''(2)}{3!}(x - 2)^3 \text{ for } x \text{ near } 2.$$

$$f(2.2) \approx 7.20 + 3.50(2.2 - 2) + \frac{1.75}{2}(2.2 - 2)^2 + \frac{0.85}{6}(2.2 - 2)^3$$

$$f(2.2) \approx 7.20 + 3.50(0.2) + \frac{1.75}{2}(0.2)^2 + \frac{0.85}{6}(0.2)^3$$

Keep in mind, that it is always best to center the polynomial closer to the x-value "asked for". That is, do not center this polynomial about $x = 0$.

The correct choice is (E).

38. Find $T(0.5) = 0.479167$ and $f(0.5) = \sin(0.5) = 0.479426$. The difference between these two numbers is $E| = |0.479426 - 0.479167| = 0.000259$.

The correct choice is (B).

39. Begin by graphing this function in the window $[0,1]$ by $[-2,2]$. Then change XMAX to 0.1, then 0.01, then 0.001, etc. Each time you will see a different graph! As x decreases from 1 to zero, $\ln x$ decreases from zero to $-\infty$. The $\sin(\ln x)$ acts like $\sin x$ as x goes from zero to $-\infty$, only it does it in the space from one to zero (like a compressed spring). The zeros of the function occur when $\sin(\ln x) = 0$ which is when $\ln x = \pm k\pi$, $k = 0,1,2,3,...$, or when $x = e^{\pm k\pi}$, $k = 0,1,2,3,...$, All of the negative powers are in the given interval, therefore there are an infinite number of zeros. HOWEVER, since each zero is over an order of magnitude larger than the next smaller zero, it is impossible to "see" more than three roots in any graphing window. Be aware of the limitations of the graphing calculator. Therefore, there are more than 4 relative extreme values for this function.

The correct choice is (E).

40. If $f^{-1}(x) = g(x)$, then $g'(x) = \dfrac{1}{f'(g(x))}$ and $g'(0) = \dfrac{1}{f'(g(0))}$.

    Solve on a graphing calculator $x^3 - 7x^2 + 25x - 39 = 0$. Thus $x = 3$.

    Since $f(3) = 0$, $g(0) = 3$.

    $f'(x) = 3x^2 - 14x + 25$, $f'(3) = 3(3^2) - 14(3) + 25 = 10$

    $g'(0) = \dfrac{1}{f'(3)} = \dfrac{1}{10}$

    The correct choice is (C).

41. Since the variables cannot be separated, use Euler's method to approximate the solution:

    When $x = 1$ and $y = 3$, $\dfrac{dy}{dx} = 1(3) - 3^2 = -6$.

    Therefore, the approximate change in y corresponding to a change in x of one (from $x = 1$ to $x = 2$) is -6.

    Hence, $y(2) \approx 3 + (-6) = -3$.

    The correct choice is (A).

42. The area A of the rectangle may be represented by $A = 2x\sqrt{64 - x^2}$ where $2x$ is the base of the rectangle and its height is $y = \sqrt{64 - x^2}$.

    $A' = (2x)\left(\dfrac{-x}{\sqrt{64 - x^2}}\right) + 2\sqrt{64 - x^2} = \dfrac{-2x^2 + 2(64 - x^2)}{\sqrt{64 - x^2}} = \dfrac{128 - 4x^2}{\sqrt{64 - x^2}}$

    Reject the critical values of $x = \pm 8$ from the denominator, since the rectangle's height would equal zero.

    The only critical values to consider are when $128 - 4x^2 = 0 \Rightarrow x = \pm\sqrt{32}$.

    Reject $x = -\sqrt{32}$ for a length and choose $x = \sqrt{32}$ as the maximum. Since A' switches from $+$ to $-$ at $x = \sqrt{32}$.

    Therefore, the maximum area is $A = 2(\sqrt{32})(\sqrt{32}) = 64$.

    The correct choice is (E).

43. If f is differentiable at $x = 0$, then it is continuous at $x = 0$.

    If f is to be continuous at $x = 0$, then $f(0) = \lim\limits_{x \to 0} f(x)$.

    $f(0) = b$; $\lim\limits_{x \to 0^+} f(x) = b$ and $\lim\limits_{x \to 0^-} f(x) = 3 \Rightarrow b = 3$.

    $$f'(x) = \begin{cases} -e^{-x}, & \text{for } x < 0 \\ a, & \text{for } x \geq 0 \end{cases}$$

    If f is differentiable at $x = 0$, then $\lim\limits_{x \to 0^+} f'(x) = \lim\limits_{x \to 0^-} f'(x)$.

    $\lim\limits_{x \to 0^+} f'(x) = a$ and $\lim\limits_{x \to 0^-} f'(x) = -1 \Rightarrow a = -1$.

    Therefore $a = -1$ and $b = 3$, or $a + b = 2$.

    The correct choice is (D).

44. Solve the differential equation by separating the variables and then using the method of partial fractions on the right side:

    $$\frac{dy}{y} = \frac{2\,dt}{t(t+2)}$$

    $$\frac{A}{t} + \frac{B}{t+2} = \frac{A(t+2) + Bt}{t(t+2)} = \frac{2}{t(t+2)}$$

    $$A(t+2) + Bt = 2$$

    Substituting $t = -2 \Rightarrow B = -1$; substituting $t = 0 \Rightarrow A = 1$

    $$\int \frac{dy}{y} = \int \frac{1}{t}\,dt + \int \frac{-1}{t+2}\,dt$$

    $$\ln|y| = \ln|t| - \ln|t+2| + \ln C;$$

    $$\ln|y| = \ln\left|\frac{Ct}{t+2}\right|;$$

    $$|y| = \left|\frac{Ct}{t+2}\right|$$

    $$y = C\left(\frac{t}{t+2}\right)$$

    When $t = 1$, $y = 1$ (initial condition) $\Rightarrow 1 = C \cdot \frac{1}{3}$, or $C = 3$

    The solution to the differential equation is $y = \dfrac{3t}{t+2}$.

    Therefore, when $t = 2$, $y = \dfrac{3(2)}{2+2} = \dfrac{3}{2}$

    The correct choice is (E).

45. The function f is a parabola with a maximum point at $(2,2)$. When $b = 2$ the horizontal line $y = 2$ is tangent at $(2,2)$. As b increases from 2, the tangent lines will move to the right of the maximum point. The "last" line tangent in the first quadrant will be just before the zero of f at $x = 3$. Find the equation of the tangent line at $x = 3$. The slope is $f'(x) = -4(x-2)$ and $f'(3) = -4$. The tangent line at this point is $y = -4(x-3)$ or $y = -4x + 12$. Thus b must be between 2 and 12. (Note that $b \neq 12$ on the technicality that the axis is not in the first quadrant.)

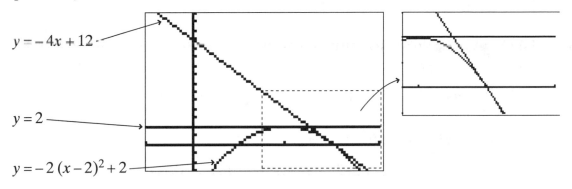

$$y = -4x + 12$$
$$y = 2$$
$$y = -2(x-2)^2 + 2$$

The correct choice is (C).

1a. The acceleration, $a(t)$, is the derivative of the velocity $v(t)$.

$$a(t) = v'(t) = -\frac{1}{t^2} \cos\left(\frac{1}{t}\right)$$

1b. The greatest velocity value occurs when $v'(t) = a(t) = 0$. This occurs when

$$\cos\left(\frac{1}{t}\right) = 0 \text{ or when } \frac{1}{t} = \frac{\pi}{2}, \frac{1}{t} = \frac{3\pi}{2}, \frac{1}{t} = \frac{5\pi}{2}, \text{etc.}$$

$$t = \frac{2}{\pi}, \frac{2}{3\pi}, \frac{2}{5\pi}, \dots$$

In the interval $0.1 < t < 0.3$, $v'(t) = a(t)$ is negative for $0.1273 < t < 0.2122$ and $a(t)$ is positive for $0.1 < t < 0.1273$ and $0.2122 < t < 0.3$. Therefore, the maximum occurs at $t = \frac{2}{5\pi} \approx 0.1273$ or near the left endpoint of the given open interval.

$$v\left(\frac{2}{5\pi}\right) \text{ or } v(0.1273) = 1$$

$$v(0.3) \approx -0.191; v(0.1) \approx -0.554$$

The maximum occurs at $t = \frac{2}{5\pi}$ or 0.1273.

1c. The position function equals $\int v(t)\, dt$. By the Fundamental Theorem of Calculus,

$$\int_{0.16}^{0.25} v(t)\, dt = x(0.25) - x(0.16)$$
$$-0.0707655 = x(0.25) - 0$$

Therefore, $x(0.25) = x(0.25) - 0$

1d.

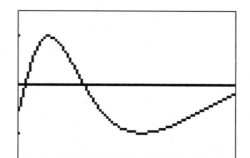

From the graph of $v(t)$, above, the velocity is below the x-axis (negative) much more than it is above, indicating that the particle moves more to the left than to the right. The particle spends more time moving to the left.

2a. $\int_0^9 \dfrac{5000}{x^3 + 50}\, dx \approx 415.420$ or 415.421 cubic meters

2b. $S(6) - R(6) = \dfrac{5000}{6^3 + 50} - 23.9665\sqrt{6} = -39.908$ or -39.909 cubic meters per hour. This means that the amount of sand in the bin is decreasing at the rate of 39.908 (or 39.909) cubic meters per hour when $t = 6$.

2c. The amount of sand, $A(t)$, in the bin is given by $A(t) = \int_0^t S(x)\, dx - \int_1^t R(x)\, dx$. The maximum occurs when $A'(t) = S(t) - R(t) = 0$ or when $S(t) = R(t)$.

OR: Before the time when $S(t) = R(t)$ the rate at which sand is being poured into the bin is greater than the rate at which it is being taken out of the bin. After this point the rate at which it is taken out is greater than the rate at which it is pouring into the bin. Therefore the maximum must occur when $S(t) = R(t)$.

2d. From part (a) the amount of sand put into the bin in the course of the day is about 415.42091 cubic meters. The amount that is taken out is $\int_1^9 23.9665\sqrt{t}\, dt \approx 415.41933$ cubic meters. Therefore, at the end of the day, the bin has 0.00158 or 0.002 cubic meters of sand.

<u>**REMINDER**</u>: **Numerical and algebraic answers need not be simplified, and if simplified incorrectly credit will be lost.**

3a. Solve the equation $\sin(x) = \frac{1}{2}$ to determine where the two graphs intersect. The values are $x = \frac{\pi}{6}$ and $x = \frac{5\pi}{6}$.

$$\text{Area} = \int_{\frac{\pi}{6}}^{\frac{5\pi}{6}} \left(\sin(x) - \frac{1}{2}\right) dx = \left(-\cos(x) - \frac{1}{2}x\right)\Big|_{\frac{\pi}{6}}^{\frac{5\pi}{6}}$$

$$= -\cos\left(\frac{5\pi}{6}\right) - \frac{1}{2}\left(\frac{5\pi}{6}\right) - \left(-\cos\left(\frac{\pi}{6}\right) - \frac{1}{2}\left(\frac{\pi}{6}\right)\right)$$

$$= -\left(-\frac{\sqrt{3}}{2}\right) - \frac{5\pi}{12} + \frac{\sqrt{3}}{2} + \frac{\pi}{12}$$

$$= \sqrt{3} - \frac{\pi}{3}$$

3b. The thickness of the washer is dx. The outer radius is $\sin(x)$ and the inner radius is $\frac{1}{2}$.

$$\text{Volume} = \pi \int_{\frac{\pi}{6}}^{\frac{5\pi}{6}} \left((\sin(x))^2 - \left(\frac{1}{2}\right)^2\right) dx$$

3c. The thickness is dx. The diameter of the semi-circles is the length of the chord between the graphs, which is $\sin(x) - \frac{1}{2}$, so the radius of the semi-circles is $\dfrac{\sin(x) - \frac{1}{2}}{2}$.

$$\text{Volume} = \frac{\pi}{2} \int_{\frac{\pi}{6}}^{\frac{5\pi}{6}} \left(\frac{\sin(x) - \frac{1}{2}}{2}\right)^2 dx$$

4a. $\dfrac{dy}{dx} = \dfrac{dy/dt}{dx/dt} = \dfrac{6\sin^2 t \cos t}{-6\cos^2 t \sin t} = -\tan t$

4b. When $t = 0$, the coordinates are $(2,0)$, which is point A. At this point, the slope is

$\dfrac{dy}{dx} = -\tan 0 = 0$ which is the same as the slope of the x-axis.

When $t = \dfrac{\pi}{2}$, the coordinates are $(0,2)$ which is point B. At this point, $\tan \dfrac{\pi}{2}$ is undefined as is the slope of the y-axis.

Thus, the slopes of the curved road and the two highways are the same at both A and B.

4c. The length of the curve is given by:

$$L = \int_a^b \sqrt{\left(\dfrac{dx}{dt}\right)^2 + \left(\dfrac{dy}{dt}\right)^2}\, dt = \int_0^{\frac{\pi}{2}} \sqrt{\left(-6\cos^2(t)\sin(t)\right)^2 + \left(6\sin^2(t)\cos(t)\right)^2}\, d$$

5a. The series is geometric with a ratio of $-\frac{1}{3}(x-2)$. Therefore it will converge when

$$\left| -\frac{1}{3}(x-2) \right| < 1$$

$$-1 < \frac{1}{3}(x-2) < 1$$

$$-3 < x - 2 < 3$$

$$-1 < x < 5$$

5b. Differentiate term by term:

$$f'(x) = -\frac{1}{3} + \frac{2}{9}(x-2) - \frac{3}{27}(x-2)^2 + \frac{4}{81}(x-2)^3 + \cdots + (-1)^n (n)\frac{(x-2)^{n-1}}{3^n} + \cdots, n \geq 1.$$

Then substitute $x = 2$ to find $f'(2) = -\frac{1}{3}$.

5c. The tangent line has a slope of $-\frac{1}{3}$ and contains the point $(2, f(2)) = (2,1)$. Its equation is

$y - 1 = -\frac{1}{3}(x-2)$ or $y = 1 - \frac{1}{3}(x-2)$. Notice that this is the constant and linear term of

the original power series. Alternatively, $f'(2)$ is the coefficient of the linear term of the power

series; $f'(2) = -\frac{1}{3}$.

5d. Differentiating the answer to (b) gives

$$f''(x) = \frac{2}{9} - \frac{6}{27}(x-2) + \cdots$$
$$f''(2) = \frac{2}{9}$$

Since the second derivative is positive at $x = 2$, the graph is concave up there and the tangent line lies below the graph of f.

6a. $g(4.5) = 13.5$. This is the area under the graph of f in the first quadrant between $x = 0$ and $x = 4.5$.

$g'(4.5) = f(4.5) = 0$; $g''(4.5) = f'(4.5) = -2$, the slope of the line segment containing $(4.5, 0)$.

6b. Average value of f equals

$$\frac{1}{5-(-3)}\int_{-3}^{5} f(x)\ dx = \frac{1}{8}\left[\int_{-3}^{3} f(x)\ dx + \int_{3}^{5} f(x)\ dx\right]$$
$$= \frac{1}{8}[27 + 2] = \frac{29}{8}$$

The two integrals are evaluated by finding the signed areas under the graph of f.

6c. There is a point of inflection where f, the derivative of g, has a maximum or minimum. This occurs only at $x = 5$, where f changes from decreasing to increasing, or where f' changes from negative to positive.

6d. The function g increases from $x = -3$ to $x = 4.5$, then decreases until $x = 7$, and then increases again until $x = 9$. There is a maximum at $(4.5, 13.5)$ (from part (a)). There is an endpoint maximum at $x = 9$. Use the signed area to find $g(9)$:

$$g(9) = g(4.5) + \int_{4.5}^{9} f(x)\ dx$$
$$= 13.5 - 0.25$$

The endpoint maximum occurs at $(9, 13.25)$.

Sample Examination IV

1. Treating y as an accumulation function, $\int_2^x f(t)\, dt$ gives the accumulated amount from $t = 2$ to $t = x$. Adding to this the amount of 4 already present at $x = 2$ gives $4 + \int_2^x f(t)\, dt$.

   Alternately $y(x) - y(2) = \int_2^x f(t)\, dt$ so $y(x) = 4 + \int_2^x f(t)\, dt$.

   The correct choice is (B).

2. The derivative of the expression is $24x^5 - 24x^2 = 24x^2(x^3 - 1) = 24x^2(x - 1)(x^2 + x + 1)$. The derivative is equal to 0 when $x = 0$ and $x = 1$. ($x^2 + x + 1$ is always positive). Since the derivative changes from negative to positive only at $x = 1$, this is the only relative minimum (First Derivative Test).

   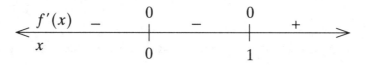

   The correct choice is (C).

3. To find the vertical tangents, set $x'(t) = 0$.

   $x'(t) = -4 \cos t$

   $x'(t) = 0$ when $-4 \cos t = 0 \Rightarrow \cos t = 0$ or $t = \dfrac{\pi}{2}, \dfrac{3\pi}{2}$.

   Vertical tangents occur at all points of the form
   $\left(x\left(\dfrac{\pi}{2}\right), y\left(\dfrac{\pi}{2}\right) \right)$ and $\left(x\left(\dfrac{3\pi}{2}\right), y\left(\dfrac{3\pi}{2}\right) \right)$.

   So the vertical tangents are at $(-1, 4)$ and $(7, 4)$.

   Note: To find the horizontal tangents, set $y'(t) = 0$.

   The correct choice is (C).

4. $\dfrac{dx}{dt} = 4t$ and $\dfrac{dy}{dt} = 3t^2$

$$\frac{dy}{dx} = \frac{dy/dt}{dx/dt} = \frac{3t^2}{4t} = \frac{3t}{4}$$

$$\frac{d^2y}{dx^2} = \frac{\dfrac{d}{dt}\left(\dfrac{dy}{dx}\right)}{\dfrac{dx}{dt}} = \frac{3/4}{4t} = \frac{3}{16t}$$

Therefore at $t = 3$, $\dfrac{d^2y}{dx^2} = \dfrac{3}{16(3)} = \dfrac{1}{16}$.

The correct choice is (A).

5. From the answer choices it is apparent that what happens at $x = \pm\frac{1}{2}$ needs to be investigated.

Substitute $x = \frac{1}{2}$: $\displaystyle\sum_{n=1}^{\infty} \frac{2^n \left(\frac{1}{2}\right)^n}{n} = \sum_{n=1}^{\infty} \frac{1}{n}$ this is the harmonic series. The harmonic series diverges. This eliminates choices (A), (B), and (D).

Now substitute $x = -\frac{1}{2}$: $\displaystyle\sum_{n=1}^{\infty} \frac{2^n \left(-\frac{1}{2}\right)^n}{n} = \sum_{n=1}^{\infty} \frac{(-1)^n}{n}$, this is the alternating harmonic series which converges. So the correct choice is (E) $-\frac{1}{2} \le x < \frac{1}{2}$

The correct choice is (E).

6. The average velocity of a particle for $t_1 \le t \le t_2$ is
$$\bar{v} = \frac{x(t_2) - x(t_1)}{t_2 - t_1}$$
where $x(t)$ is the position of the particle at time t.

The position function for this problem is given as $x(t) = \ln t$. Substituting into the expression above gives:
$$\bar{v} = \frac{\ln e - \ln 1}{e - 1} = \frac{1 - 0}{e - 1} = \frac{1}{e - 1}$$

The correct choice is (C).

7. The area is $\dfrac{1}{2}\displaystyle\int_{\alpha}^{\beta} r^2 \, d\theta$ or equivalently $\dfrac{1}{2}\displaystyle\int_{\alpha}^{\beta} [f(\theta)]^2 \, d\theta$.

The correct choice is (E).

8. Using the method of Integration by Parts with $u = 2x$, $du = 2\,dx$, $dv = \cos(2x)\,dx$, $v = \frac{1}{2}\sin(2x)$ gives

$$\int (2x) \cos(2x)\,dx = (2x)\left(\frac{1}{2}\sin(2x)\right) - \int \frac{1}{2}\sin(2x)(2\,dx)$$

$$= x\sin(2x) + \frac{1}{2}\cos(2x)$$

Or the set up could be $u = x$, $du = dx$, $dv = \cos(2x)(2\,dx)$, $v = \sin(2x)$ which gives

$$\int (2x)\cos(2x)\,dx = x\sin(2x) - \int \sin(2x)\,dx$$

$$= x\sin(2x) + \frac{1}{2}\cos(2x)$$

The correct choice is (D).

9. $|x + 2| = \begin{cases} x + 2, & x \geq -2 \\ -x - 2, & x < -2 \end{cases}$

$$\int_{-3}^{3} |x + 2|\,dx = \int_{-3}^{-2}(-x - 2)\,dx + \int_{-2}^{3}(x + 2)\,dx$$

$$= \left(\frac{-x^2}{2} - 2x\right)\Big|_{-3}^{-2} + \left(\frac{x^2}{2} + 2x\right)\Big|_{-2}^{3}$$

$$= \left[(-2 + 4) - \left(-\frac{9}{2} + 6\right)\right] + \left[\left(\frac{9}{2} + 6\right) - (2 - 4)\right]$$

$$= \left(2 + \frac{9}{2} - 6\right) + \left(\frac{9}{2} + 6 - 2 + 4\right)$$

$$= 13$$

Alternate Solution:

Sketch the graph of $y = |x + 2|$ for $-3 \leq x \leq 3$ and find the areas of the two triangles with bases on the x-axis.

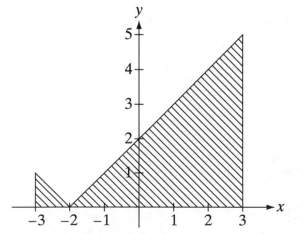

The correct choice is (C).

10.

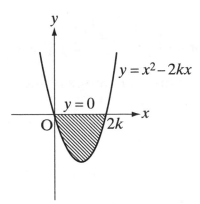

The graph of $y = x^2 - 2kx$ is a parabola whose x-intercepts are $x = 0$ and $x = 2k$.

The x-intercepts are found by setting $y = 0$ and solving for x.

$$x^2 - 2kx = 0$$

$$x(x - 2k) = 0$$

$$x = 0, x = 2k$$

The region R is shown shaded in the figure above. <u>Note</u>: The area of region R is bounded by two functions, $y = 0$ and $y = x^2 - 2kx$. When $0 \leq x \leq 2k$ the function $y = 0 \geq y = x^2 - 2kx$.

Since the area of the region is given as 36,

$$\int_0^{2k} 0 - (x^2 - 2kx)\, dx = 36$$

$$\int_0^{2k} (-x^2 + 2kx)\, dx = 36$$

$$\left. \frac{-x^3}{3} + \frac{2kx^2}{2} \right|_0^{2k} = 36$$

$$\left. \frac{-x^3}{3} + kx^2 \right|_0^{2k} = 36$$

$$\frac{-8k^3}{3} + 4k^3 = 36$$

$$4k^3 = 108$$

$$k^3 = 27$$

$$k = 3$$

The correct choice is (B).

11. The given problem is equivalent to $\lim\limits_{h \to 0} \dfrac{\displaystyle\int_0^h \frac{\sin^2 t}{t^2}\, dt}{h}$ which is of the form $\dfrac{0}{0}$.

Using L'Hôpital's Rule, take the first derivatives of the numerator (by the Fundamental Theorem of Calculus) and denominator,

$$\lim_{h \to 0} \frac{\displaystyle\int_0^h \frac{\sin^2 t}{t^2}\, dt}{h} = \lim_{h \to 0} \frac{\frac{\sin^2 h}{h^2}}{1}$$

$$= \lim_{h \to 0} \frac{\sin^2 h}{h^2} = \lim_{h \to 0} \left(\frac{\sin h}{h} \cdot \frac{\sin h}{h} \right) = 1$$

Keep in mind that $\lim\limits_{h \to 0} \dfrac{\sin h}{h} = 1$.

The correct choice is (C).

12. Since $f(x)$ is continuous on $[-3, 7]$ and differentiable on $(-3, 7)$, the Mean Value Theorem guarantees a point c, $-3 < c < 7$, such that $f'(c) = \dfrac{f(7) - f(-3)}{7 - (-3)} = \dfrac{2 - 4}{10} = -\dfrac{1}{5}$

The correct choice is (D).

13. By the Chain Rule, $h'(x) = f'(g(x))\, g'(x)$.

Thus $h'(2) = f'(g(2))\, g'(2)$.

Then since $f'(x) = 2x + 3$ and $g(2) = 1$, $f'(g(2)) = 2(1) + 3 = 5$ and from the table $g'(2) = 3$. Therefore, $h'(2) = 5(3) = 15$

The correct choice is (D).

14. Write $g(x)$ as $g(x) = [f(x)]^{-1}$, then differentiate using the Power Rule and the Chain Rule. $g'(x) = (-1)[f(x)]^{-2} f'(x)$. Then substitute $x = 2$ and use the values from the table,

$$g'(2) = (-1)[f(2)]^{-2} f'(2) = (-1)(-8)^{-2}(-4) = \frac{4}{64} = \frac{1}{16}$$

<u>Note</u>: We could have used the Quotient Rule to find $g'(x)$.

$$g'(x) = \frac{-f'(x)}{[f(x)]^2}; \text{ so } g'(2) = \frac{-f'(2)}{[f(2)]^2} = \frac{4}{64} = \frac{1}{16}$$

The correct choice is (C).

15. I. $\sum\limits_{n=1}^{\infty} \left(1 - \frac{4}{3}\right)^n$ is a geometric series whose common ratio is $\left(-\frac{1}{3}\right)$ and will <u>converge</u>.

   II. $\sum\limits_{n=1}^{\infty} \left(1 + \frac{4}{3}\right)^n$ is also a geometric series whose ratio is $\frac{7}{3}$ and will <u>diverge</u>.

   For III, $\lim\limits_{n \to \infty} \left(1 + \frac{1}{n}\right)^n = e$ and therefore it will <u>diverge</u> by the nth-Term Test for Divergence.

   The correct choice is (A).

16. This could be done by differentiating the given expression three times. However, the general term of the Taylor series is $\dfrac{f^{(n)}(a)(x-a)^n}{n!}$.

   The third derivative ($n = 3$) at $x = a$ is found in the fourth term of $f(x)$; it is the coefficient of $\dfrac{(x-3)^3}{3!}$. Thus $f'''(3) = 7$.

   The correct choice is (E).

17. Substituting for $f(x)$ gives:

$$\int_2^2 [f(x) - g(x)]\, dx = \int_2^2 [15 - g(x) - g(x)]\, dx$$
$$= \int_{-2}^2 [15 - 2g(x)]\, dx$$
$$= 15\int_{-2}^2 dx - 2\int_{-2}^2 g(x)\, dx$$
$$= 15x\big|_{-2}^2 - 2\int_{-2}^2 g(x)\, dx$$
$$= 60 - 2\int_{-2}^2 g(x)\, dx$$

   <u>Note</u>: Since there is no information about the symmetry of $g(x)$, it cannot be said that $\int_{-2}^2 g(x)\, dx = 2\int_0^2 g(x)\, dx$. Thus choice (D) is not correct.

   The correct choice is (E).

18. $\int_a^x \frac{d}{dt}[f(t)]\, dt = f(t)\Big|_a^x = f(x) - f(a)$

$\int_2^4 \left[\frac{d}{dt}(3t^2 + 2t - 1)\right] dt = \int_2^4 (6t + 2)\, dt = 3t^2 + 2t\Big|_2^4 = (48 + 8) - (12 + 4) = 40$

Note: Do not confuse $\int_a^x \frac{d}{dt}[f(t)]\, dt$ with $\frac{d}{dx}\left[\int_a^x f(t)\, dt\right]$.

$\int_a^x \frac{d}{dt}[f(t)]\, dt = f(x) - f(a)$ while $\frac{d}{dx}\left[\int_a^x f(t)\, dt\right] = f(x)$.

The correct choice is (B).

19. The comparison test shows that I is true since $a_n < b_n$. II could be true, but there is no convincing reason to believe it must be true. III is true since the sum of two convergent series converges.

The correct choice is (D).

20. Since each part of the definition of the function is continuous, check the place where they join, at $x = 5$. Here, both parts reduce to $1 + e^{-5}$. Therefore, the function is continuous, so I is true.

$f'(x) = \begin{cases} -e^{-x} & \text{for } 0 \le x \le 5 \\ e^{x-10} & \text{for } 5 < x \le 10 \end{cases}$

The derivative on the left side of 5 is negative and on the right side is positive. It is never zero so the derivative is not continuous, and II is false.

$f''(x) = \begin{cases} e^{-x} & \text{for } 0 \le x \le 5 \\ e^{x-10} & \text{for } 5 < x \le 10 \end{cases}$

The second derivative is always positive, so the graph of $f(x)$ is concave up everywhere, and III is true.

The correct choice is (D).

21. The equation of the circle is $x^2 + y^2 = 9$. The area of an equilateral triangle of side length s is $\frac{\sqrt{3}}{4} s^2$.

The cross section of the solid is an equilateral triangle of side $2y$ and area of $\sqrt{3} y^2$ or $\sqrt{3}(9 - x^2)$.

The Volume $V = \sqrt{3} \int_{-3}^{3} (9 - x^2)\, dx$

$$= \sqrt{3} \left(9x - \frac{x^3}{3}\right)\Big|_{-3}^{3}$$

$$= \sqrt{3}(27 - 9) - \sqrt{3}(-27 + 9)$$

$$= 36\sqrt{3}$$

The correct choice is (E).

22. $R(t)$, the rate of change of the velocity, is the acceleration. $R(0) = a(0) = -\sqrt{3}$. The velocity $v(0) = 1$ (given). Since the velocity is positive, the particle is moving to the right. Since the velocity and accleration have different signs, the speed is decreasing.

The correct choice is (D).

23. Cancelling the 8^n gives $s_n - \left(\frac{1}{8^2}\right)\left(\frac{(8-n)^{200}}{(3-n^2)^{100}}\right)$.

If the binomials in the second fraction are expanded, both numerator and denominator will be polynomials with leading (highest power) terms of $(+1)\, n^{200}$. The limit of this part of the expression is $(+1)$, so the limit of the entire expression is $\frac{1}{64}$.

The correct choice is (D).

24. In general, if $y = \arcsin(u)$, then $y' = \dfrac{du}{\sqrt{1 - u^2}}$.

In this problem, $y = \arcsin\left(\dfrac{3x}{4}\right)$ and let $u = \dfrac{3x}{4}$

$$
\begin{aligned}
y' &= \frac{\frac{3}{4}}{\sqrt{1 - \left(\frac{3x}{4}\right)^2}} \\
&= \frac{\frac{3}{4}}{\sqrt{1 - \frac{9x^2}{16}}} \\
&= \frac{\frac{3}{4}}{\sqrt{\frac{16 - 9x^2}{16}}} \\
&= \frac{\frac{3}{4}}{\frac{1}{4}\sqrt{16 - 9x^2}} \\
&= \frac{3}{\sqrt{16 - 9x^2}}
\end{aligned}
$$

The correct choice is (C).

25. This question may be answered without solving the differential equation. For horizontal asymptotes $\lim\limits_{x \to \pm\infty} \dfrac{dy}{dx} = 0$. The derivative will be zero when $y = 0$ and when $\left(8 - \dfrac{y}{1,000}\right) = 0$ or $y = 8,000$.

Or solving gives $y = \dfrac{8000e^{8x}}{e^{8x} + C}$. $\lim\limits_{x \to +\infty} \dfrac{8000e^{8x}}{e^{8x} + C} = 8,000$ and $\lim\limits_{x \to -\infty} \dfrac{8000e^{8x}}{e^{8x} + C} = 0$

The correct choice is (E).

26. The integral is improper since the denominator of the integrand is zero at $x = 4$.

Break up the integral into $\displaystyle\int_2^4 \dfrac{dx}{x - 4} + \int_4^6 \dfrac{dx}{x - 4}$.

If either integral diverges, then the original integral also diverges.

The indefinite integral $\displaystyle\int \dfrac{dx}{x - 4} = \ln|x - 4| + C$.

The first improper integral diverges since it is defined to be

$$\lim_{t \to 4^-} \int_2^t \dfrac{dx}{x - 4} = \lim_{t \to 4^-} (\ln|x - 4| - \ln|2 - 4|) = -\infty$$

Since one of these improper integrals is divergent, the entire integral $\displaystyle\int_2^6 \dfrac{dx}{x - 4}$ diverges.

The correct choice is (E).

27. The slope field gives the slope of the solution of the differential equation. This function increases, levels off, and then increases again. So, I is false; and, II and III are true.

    The correct choice is (D).

28. At point A, the function is decreasing indicating $f' < 0$ and concave up indicating that $f'' > 0$, thus $f' < f''$ is true at A.

    At point B, the function is increasing indicating $f' > 0$ and concave down indicating that $f'' < 0$, thus $f' < f''$ is false at B.

    At point C, the function has a horizontal tangent indicating $f' = 0$ and concave down indicating that $f'' < 0$, thus $f' < f''$ is false at point C.

    Hence, $f'(x) < f''(x)$ at point A only.

    The correct choice is (A).

29. Method I:

    To find $\dfrac{dy}{dx}$, use implicit differentiation,

    $$e^{xy} = 2$$
    $$(e^{xy})\left(x\,\frac{dy}{dx} + y\right) = 0$$
    $$xe^{xy}\,\frac{dy}{dx} + ye^{xy} = 0$$
    $$\frac{dy}{dx} = \left.\frac{-ye^{xy}}{xe^{xy}}\right|_{(1,\ln 2)} = \frac{(-\ln 2)(e^{\ln 2})}{e^{\ln 2}} = -\ln 2$$

    Method II:

    Alternate Solution:

    $$\ln e^{xy} = \ln 2$$
    $$xy = \ln 2$$
    $$y = \frac{\ln 2}{x}$$
    $$\frac{dy}{dx} = \frac{-\ln 2}{x^2}$$
    $$\left.\frac{dy}{dx}\right|_{x=1} = -\ln 2$$

    The correct choice is (A).

30. Speed $= |x'(t)| = |4t^3 - 30t^2 + 58t - 36|$. The greatest speed may be found by using a calculator to graph. Otherwise evaluate, $4t^3 - 30t^2 + 58t - 36|$ for each choice. Of the values given, $t = 4$ gives the greatest speed. (Note: the greatest speed in the interval is at $t \approx 3.690$).

    The correct choice is (D).

31. First, observe that since $f'(x) > 0$ for all x, $f(x)$ is increasing. That fact eliminates the graphs of choices (A), (B), and (C). When the first derivative is increasing for $x \le 0$, the second derivative will be positive and function f will be concave upwards. Likewise, when the derivative is decreasing for $x \ge 0$ the function f will be concave downward. This describes graph (E) and eliminates graph (D).

    The correct choice is (E).

32. Integrate the velocity vector to find the position vector: $s(t) = \langle -2 \cos t + C_1, 3 \sin t + C_2 \rangle$.

    Substitute $t = 0$ to evaluate the two constants, C_1 and C_2:

    $$-2 \cos(0) + C_1 = 1 \Rightarrow C_1 = 3 \text{ and } 3 \sin(0) + C_2 = 1 \Rightarrow C_2 = 1$$

    So the position vector is $\langle -2 \cos t + 3, 3 \sin t + 1 \rangle$. Substitute $t = 2$ to find the position vector:

    $$\langle -2 \cos(2) + 3, 3 \sin(2) + 1 \rangle = \langle 3.832, 3.728 \rangle.$$

    The correct choice is (A).

33. By the Fundamental Theorem of Calculus $f'(t) = e^{t \cos(t)} (\cos(t) - t \sin(t))$. The maximum of f will occur where its derivative changes from positive to negative or at the endpoints of the domain. Graph the derivative in a convenient window, such as x [0,10] by y [–10,10]

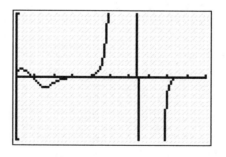

    Note that this graph does NOT have a vertical asymptote – this should be obvious from the equation of the derivative.

However it is best to use ZoomFit x [0,10] by y [-922, 854]

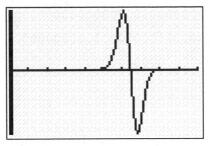

The derivative changes from positive to negative to the left of $t = 1$ ($t = 0.860$) and just past 6 ($t = 6.437$). The other zeros are 3.426, and 9.529. Based on the size of the area of the regions between the graph of $f'(t)$ and the horizontal axis, $t = 6.437$ must be the location of the absolute maximum.

By looking at the answer choices it is *not* necessary to actually calculate the zeros, since only one choice is between 6 and 7.

Alternate Solution:

The integral is of the form $\int e^u \, du$ where $u = x \cos(x)$ and $du = (\cos(x) - x \sin(x)) \, dx$. Therefore $f(t) = e^{x \cdot \cos x}\big|_0^t = e^{t \cos(t)} - 1$. Graph this and observe that the maximum is between $t = 6$ and $t = 7$.

The correct choice is (C).

34. $y' = 3x^2$; Arc length $= \int_a^b \sqrt{1 + \left(\dfrac{dy}{dx}\right)^2} \, dx = \int_0^1 \sqrt{1 + 9x^4} \, dx = 1.548$

The correct choice is (D).

35. Graph $f'(x)$ in the viewing window $x[0,12]$ by $y[-2,2]$. In the given interval, the derivative changes from positive to negative twice, indicating that $f(x)$ will have two relative maxima (First Derivative Test).

The correct choice is (C).

36. Distance $= \int_{t_1}^{t_2} v(t)\, dt = \int_{t_1}^{t_2} 2e^{2t}\, dt$.

Find the values of t_1 and t_2 when the velocity is 2 and 4.

$$\begin{cases} 2e^{2t} = 2 \Rightarrow e^{2t} = 1 \text{ or } t = 0 \\ 2e^{2t} = 4 \Rightarrow e^{2t} = 2 \text{ or } t = \frac{1}{2}\ln 2 \end{cases}$$

Thus the distance the particle travels equals $\int_{0}^{\frac{1}{2}\ln 2} 2e^{2t}\, dt = 1$.

The correct choice is (A).

37. The area of the square is 4. The area under the parabola $y = -x^2 + 2x$ is given by

$$\int_{0}^{2} (-x^2 + 2x)\, dx = -\frac{x^3}{3} + x^2 \Big|_{0}^{2} = \frac{4}{3}$$

The area above the parabola is $4 - \left(\frac{4}{3}\right) = \frac{8}{3}$.

The probability is the ratio of the area above the parabola to the total area: $\dfrac{\frac{8}{3}}{4} = \dfrac{2}{3}$.

The correct choice is (C).

38. The second derivative, $f''(x) = 5(x^2 + 5)^4 (2x)$, changes sign only once, at $x = 0$. The other factors are all positive for all values of x. Therefore, there is only one point of inflection.

The correct choice is (B).

39. Begin by finding the antiderivative term by term. Note the constant of integration:

$$f(x) = \int \left(3x - \frac{9x^3}{2} + \frac{81x^5}{40} - \frac{3x^7}{60} + \cdots \right) dx$$

$$= C + \frac{3x^2}{2} - \frac{9x^4}{2 \cdot 4} + \frac{81x^6}{40 \cdot 6} - \frac{3x^8}{60 \cdot 8} + \cdots$$

$$= C + \frac{3x^2}{2} - \frac{9x^4}{8} + \frac{27x^6}{80} - \frac{3x^8}{480} + \cdots$$

Using the initial condition $f(0) = 2$ and find C:

$$2 = C + \frac{3(0)^2}{2} - \frac{9(0)^4}{8} + \frac{27(0)^6}{80} - \frac{3(0)^8}{480} + \cdots$$

Therefore, $C = 2$ and

$$f(x) = 2 + \frac{3x^2}{2} - \frac{9x^4}{8} + \frac{27x^6}{80} - \frac{3x^8}{480} + \cdots$$

The correct choice is (E).

40. Since $f(x+h) - f(x) = 4xh + 2h^2$, so $\dfrac{f(x+h) - f(x)}{h} = 4x + 2h$.

    $$f'(x) = \lim_{h \to 0} \frac{f(x+h) - f(x)}{h} = 4x$$

    So $f(x) = \displaystyle\int 4x\,dx - 2x^2 + C$

    Since $f(2) = 5$, so $5 = 8 + C$ or $C = -3$.

    Therefore, $f(x) = 2x^2 - 3$ and $f(1) = -1$.

    The correct choice is (B).

41. Use the method of separation of variables to solve the differential equation:

    $$\frac{dy}{y} = e^x dx$$

    $$\ln|y| = e^x + C_1$$

    $$|y| = e^{e^x + C_1}$$

    $$y = Ce^{e^x}$$

    Then substitute the initial condition ($y = 3$ when $x = 0$) to find C:

    $$3 = Ce^{e^0}$$
    $$3 = Ce$$
    $$\frac{3}{e} = C$$

    Therefore, $y = \dfrac{3}{e}e^{e^x} = \dfrac{3e^{e^x}}{e}$

    The correct choice is (E).

42. <u>Method I</u>: By the Fundamental Theorem of Calculus

$g'(x) = (5 + 4x - x^2)(2^{-x}) = -(x - 5)(x + 1)(2^{-x})$ On the interval $(3,5)$, $g' > 0$ so g is increasing $(3,5)$ on, so I is true. On the interval $(5,7)$, $g' < 0$ so g is decreasing on $(5,7)$, so II is false. Use a graphing calculator to find $g(7) = \int_3^7 (5 + 4t - t^2)(2^{-t})\, dt = 0.562$, so III is false.

<u>Method II</u>: Graph the integrand $(5 + 4x - x^2)(2^{-x})$ in a suitable window such as $[3,7]$ by $[-1,2]$ (See figure). This is the derivative of g and from the graph it is clear that the derivative is positive on $(3,5)$ indicating g is increasing, so I is true.

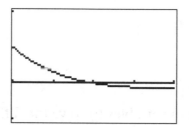

The derivative is negative on $(5,7)$ indicating that g is decreasing, so II is false.

$g(7)$ is the net area between the axis and the graph. Since there is more area above the axis than below, $g(7) > 0$ and III is false.

The correct choice is (A).

43. Using the disk method, the volume is

$$\pi k^2 \int_0^5 (x - 5)^4 dx = 2500\pi$$

Integrating and solving for k,

$$\pi k^2 \frac{(x - 5)^5}{5}\Big|_0^5 = 2500\pi$$

$$\pi k^2 (5^4) = 2500\pi$$

$$k^2 = 4 \text{ or } k = 2$$

The correct choice is (E).

44. a is a constant equal to $f(2)$ which is negative, as seen on the graph.

b will correspond to $f'(2) = 0$, since at $x = 2$, f has a relative minimum.

c corresponds to $\dfrac{f''(2)}{2!}$ which is greater than 0, since $f(x)$ is concave upward at $x = 2$.

<u>Alternate Solution</u>: The Taylor polynomial is a parabola whose vertex is at the point $(2, a)$ which is also the relative minimum point of the function graphed. Therefore $a < 0$, $b = 0$. Since the curve is concave upward at this point, the parabola must open upward, therefore $c > 0$.

The correct choice is (A).

45. Euler's Method, the tangent line approximation, calculates the approximate value of y as

$$y \approx f(x) + f'(3)\,\Delta x = -2 + \left(-\frac{3^2}{-2}(-0.3)\right) = -2 - 1.35 = -3.35$$

The correct choice is (E).

1a. $\lim\limits_{x \to -\infty} f(x) = 0$; $\lim\limits_{x \to \infty} f(x) = 0$. As x increases or decreases without bound, $e^{-\frac{x^2}{2}}$ approaches zero.

1b. Using the numerical derivative feature on your calculator, $f'(3) = -0.013$.

By hand, $f'(x) = \dfrac{1}{\sqrt{2\pi}} \left(e^{-\frac{x^2}{2}} \right)(-x)$

$f'(3) = \dfrac{-3}{\sqrt{2\pi}} e^{-\frac{9}{2}}$

1c. $f''(x) = \dfrac{1}{\sqrt{2\pi}} \left(-e^{-\frac{x^2}{2}} + x^2 e^{-\frac{x^2}{2}} \right)$

$= \dfrac{1}{\sqrt{2\pi}} \left(e^{-\frac{x^2}{2}} \right)(x^2 - 1)$

The second derivative changes sign at $x = 1$ and $x = -1$. These are the x-coordinates of the points of inflection.

1d. The area is $\displaystyle\int_{-1}^{1} \dfrac{1}{\sqrt{2\pi}} e^{-\frac{x^2}{2}} dx = 0.682$ or 0.683, by use of a graphing calculator.

2a. Substitute $L = 10,000$ when $d = 40$ into $L = \dfrac{k}{d^2}$ to find k:

$$10,000 = \frac{k}{40^2} \Rightarrow = 10^4 \cdot 40^2 \text{ or } 1.6 \times 10^7.$$

2b. From trigonometry, $\dfrac{40}{d} = \cos \theta$, so $d = \dfrac{40}{\cos \theta}$.

Since $L = \dfrac{k}{d^2}$ then $L(\theta) = \dfrac{(1.6)(10^7)}{\left(\frac{40}{\cos \theta}\right)^2} = 10^4 \cos^2 \theta$

2c. Differentiate $L = 10^4 \cos^2 \theta$ (from part (b)) with respect to t (time) to find the rate of change of L:

$$\frac{dL}{dt} = 2 \cdot 10^4 \cos \theta (-\sin \theta) \frac{d\theta}{dt}$$

From the given, substitute $\theta = \dfrac{\pi}{4}$ and $\dfrac{d\theta}{dt} = \dfrac{\pi}{30}$ to find the numerical value of $\dfrac{dL}{dt}$.

$$\frac{dL}{dt} = 2 \cdot 10^4 \cos\left(\frac{\pi}{4}\right)\left(-\sin \frac{\pi}{4}\right)\left(\frac{\pi}{30}\right)$$

$$= 2 \cdot 10^4 \left(\frac{\sqrt{2}}{2}\right)\left(-\frac{\sqrt{2}}{2}\right)\left(\frac{\pi}{30}\right)$$

$$- \frac{-10^4 \pi}{30} - -1047.198$$

Be careful how the answer is phrased: "The strength is decreasing at $+1047.198$ lumens/sec." or, "The strength is changing at -1047.198 lumens/sec." The negative sign indicates that it is decreasing." Do NOT write "The strength is decreasing at -1047.198 lumens/sec."

REMINDER: Numerical and algebraic answers need not be simplified, and if simplified incorrectly credit will be lost.

3a. Area $= \int_{1}^{5} \left(1 - \frac{1}{x}\right) dx = x - \ln(x)\Big|_{1}^{5}$

$$= 5 - \ln(5) - (1 - \ln(1))$$
$$= 4 - \ln(5)$$

3b. The thickness of the washer is dx. The outer radius is $2 - \frac{1}{x}$ and the inner radius is 1.

Volume $= \pi \int_{1}^{5} \left(\left(2 - \frac{1}{x}\right)^{2} - (2 - 1)^{2}\right) dx$

3c. To find the volume, integrate the given cross-section area by the thickness dx.

$\int_{1}^{5} (2^{x} - 2) dx = \frac{2^{r}}{\ln 2} - 2x\Big|_{1}^{5}$

$$= \frac{2^{5}}{\ln 2} - \frac{2^{1}}{\ln 2} - 10 - 2$$
$$= \frac{30}{\ln 2} - 12$$

4a. To solve the differential equation, first separate the variables:

$$\frac{dh}{dt} = -k\sqrt{h}$$

$$\frac{dh}{\sqrt{h}} = -k\,dt$$

Integrating both sides:

$$2\sqrt{h} = -kt + C$$

Now, use the initial condition $h(0) = 16$ to find C.

$$2\sqrt{16} = -k(0) + C \rightarrow C = 8$$

$$2\sqrt{h} = -kt + 8.$$

Now square both sides and solve for h:

$$4 \cdot h(t) = (-kt + 8)^2$$

$$h(t) = \tfrac{1}{4}(8 - kt)^2$$

4b. When $t = 8$, $h = 12.95$;

So, $12.25 = \tfrac{1}{4}(8 - 8k)^2$

Multiply both sides by 4 and take the positive square root (Note that using the negative square root gives a value of $k > 1$):

$$7 = 8 - 8k \Rightarrow k = \tfrac{1}{8} = 0.125$$

4c. The tank will be completely empty when $h(t) = 0$.

$$0 = \tfrac{1}{4}\left(8 - \tfrac{1}{8}t\right)^2$$

Solving for t,

After 64 hours, the tank will be completely empty.

5a. To find the point(s), first find the slope of chord AB and set it equal to the derivative of f:

Slope of chord AB$= \dfrac{15-(-15)}{-3-3} = -5$

$f'(x) = 4 - 3x^2$

So, $4 - 3x^2 = -5$

$3x^2 = 9$

$x^2 = 3$

$x = \pm\sqrt{3}$

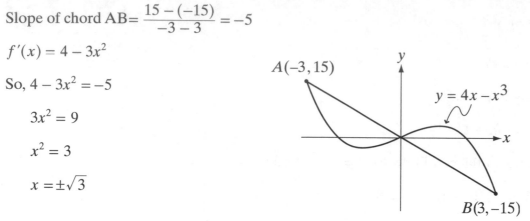

$A(-3,15)$

$y = 4x - x^3$

$B(3,-15)$

Substitute $x = \sqrt{3}$ and $x = -\sqrt{3}$ into the curve $y = 4x - x^3$ to get $(\sqrt{3},\sqrt{3})$ and $(-\sqrt{3}, -\sqrt{3})$.

The diagram shows that there are 2 points on the curve where the tangent line is parallel to chord AB.

5b. For $0 < x < 3$, $f(x)$ lies above the chord AB (see diagram above), so the vertical distance is $V(x) = (4x - x^3) - (-5x) = 9x - x^3$.

<u>Note</u>: Chord AB with coordinates $(-3,15)$ and $(3, -15)$ has an equation of $y = -5x$.

5c. To find the maximum vertical distance, differentiate $V(x)$ and find its zeros.

$V(x) = 9x - x^3$

$V'(x) = 9 - 3x^2;\ V'(x) = 0$ when $9 - 3x^2 = 0 \Rightarrow 3x^2 = 9 \Rightarrow x^2 = 3$ or $x = \sqrt{3}$

(<u>Note</u>: $x = -\sqrt{3}$ is not in the domain $0 < x < 3$)

$x = \sqrt{3}$ is a maximum since $V''(x) = -6x < 0$ at $x = \sqrt{3}$ (Second Derivative Test)

So the maximum vertical distance is $V(\sqrt{3}) = 9\sqrt{3} - (\sqrt{3})^3 = 6\sqrt{3}$

<u>Note</u>: Substitute the value of $x = \sqrt{3}$ into $V(x)$ to find the maximum vertical distance. Do not leave the answer as $x = \sqrt{3}$.

6a. The slope is $f'\left(\frac{\pi}{2}\right) = \ln\left(\sin\frac{\pi}{2}\right) = \ln(1) = 0.$ The point is $\left(\frac{\pi}{2}, 1\right)$ and an equation of the line is $y - 1 = 0\left(x - \frac{\pi}{2}\right)$ or $y = 1.$

6b. Use the Chain Rule to differentiate f': $f''(x) = \dfrac{1}{\sin x} \cos x = \cot x$

6c. Yes, the function changes concavity at $x = \frac{\pi}{2}$. To the left of this value, $\cot x > 0$, and to the right, $\cot x < 0$. Since the second derivative changes sign, you can conclude that the function changes concavity there.

6d. From the given information and the results of the previous parts you know that $f\left(\frac{\pi}{2}\right) = 1$, $f'\left(\frac{\pi}{2}\right) = 0$, and $f''\left(\frac{\pi}{2}\right) = \cot\frac{\pi}{2} = 0$. Then $f'''(x) = -\csc^2 x$ and $f'''\left(\frac{\pi}{2}\right) = -1$.

So the Taylor polynomial of degree three is

$$T_3(x) = 1 + \frac{0}{1!}\left(x - \frac{\pi}{2}\right) + \frac{0}{2!}\left(x - \frac{\pi}{2}\right)^2 - \frac{1}{3!}\left(x - \frac{\pi}{2}\right)^3$$

$$= 1 - \frac{1}{6}\left(x - \frac{\pi}{2}\right)^3$$

Sample Examination V

1. If $y = (2x^2 + 1)^4$, then $\dfrac{dy}{dx} = 4(2x^2 + 1)^3(4x) = 16x(2x^2 + 1)^3$.

   The correct choice is (E).

2. Let $u = x^2 + 1$

   $$du = 2x\,dx \Rightarrow x\,dx = \frac{1}{2}du$$

   $$\int x(x^2 + 1)^{\frac{1}{2}}dx = \frac{1}{2}\int u^{\frac{1}{2}}\,du = \frac{1}{2}\left(\frac{2}{3}u^{\frac{3}{2}}\right) = \frac{1}{3}u^{\frac{3}{2}} = \frac{1}{3}(x^2 + 1)^{\frac{3}{2}} + C$$

   The correct choice is (C).

3. The quickest way to do this problem is to simplify $f(x) = x\sqrt[3]{x} = x\left(x^{\frac{1}{3}}\right) = x^{\frac{4}{3}}$.

   Therefore, $f'(x) = \frac{4}{3}x^{\frac{1}{3}}$.

   The correct choice is (C).

4. Since $x = 2t + 3$ and $y = t^2 + 2t$, $\dfrac{dy}{dt} = 2t + 2$ and $\dfrac{dx}{dt} = 2$.

   So $\dfrac{dy}{dx} = \dfrac{dy/dt}{dx/dt} = \dfrac{2t + 2}{2} = \dfrac{2(t + 1)}{2} = t + 1$.

   At $t = 1$, $\dfrac{dy}{dx} = 2$. When $t = 1$, $x = 5$ and $y = 3$.

   So the tangent line to the curve at $t = 1$ is $y - 3 = 2(x - 5)$ or $y = 2x - 7$.

   The correct choice is (A).

5. $\displaystyle\int_0^8 \dfrac{1}{\sqrt[3]{8 - x}}\,dx$ is an improper integral.

   $$\int_0^8 \frac{1}{\sqrt[3]{8 - x}}\,dx = \lim_{b \to 8^-}\int_0^b (8 - x)^{-\frac{1}{3}}dx = \lim_{b \to 8^-}\left[-\frac{3}{2}(8 - x)^{\frac{2}{3}}\right]\Big|_0^b = -\frac{3}{2}(0) + \frac{3}{2}(4) = 6$$

   The correct choice is (C).

6. This problem is solved using integration by parts.

$$u = x \qquad dv = \sin x \, dx$$
$$du = dx \qquad v = -\cos x$$

$$\int x \sin x \, dx = uv - \int v \, du = -x \cos x - \int -\cos x \, dx$$

$$= -x \cos x + \int \cos x \, dx$$

$$= -x \cos x + \sin x + C$$

The correct choice is (D).

7. Statement I is false. Since $\int_0^3 f(x) \, dx$ equals a constant, $\dfrac{d}{dx} \int_0^3 f(x) \, dx = 0$ by the Fundamental Theorem of Calculus.

Statement II is false because $\int_3^x f'(x) \, dx = f(x) - f(3)$ (Fundamental Theorem of Calculus).

Statement III is a true representation of the Fundamental Theorem of Calculus.

The correct choice is (B).

8. To find $\dfrac{dy}{dx}$, or y', use implicit differentiation.

$$\sin(xy) = x^2$$
$$\cos(xy)(xy' + y) = 2x$$
$$xy' \cos(xy) + y \cos(xy) = 2x$$
$$xy' \cos(xy) = 2x - y \cos(xy)$$
$$y' = \frac{2x - y \cos(xy)}{x \cos(xy)}$$

Now, since $\dfrac{1}{\cos(xy)} = \sec(xy)$, $y' = \dfrac{2x \sec(xy) - y}{x}$

The correct choice is (E).

9. To be concave up the second derivative must be positive. This means that the first derivative must be increasing, which will show up in the tables as an increase in the differences between the values in the second column. This occurs only in table (B).

x	$g(x)$	Difference
1	4	
2	6	2
3	9	3
4	14	5

The correct choice is (B).

10. $\int \dfrac{dx}{x^2 + 4x} = \int \dfrac{dx}{x(x+4)}$

Decompose $\dfrac{1}{x(x+4)}$ into partial fractions:

$$\frac{1}{x(x+4)} = \frac{A}{x} + \frac{B}{x+4} = \frac{A(x+4) + Bx}{x(x+4)} = \frac{Ax + 4A + Bx}{x(x+4)} = \frac{x(A+B) + 4A}{x(x+4)}$$

Since $\dfrac{1}{x(x+4)} = \dfrac{x(A+B) + 4A}{x(x+4)}$, equate the coefficients in the numerators and

consequently $4A = 1$ and $A + B = 0 \Rightarrow A = \dfrac{1}{4}$ and $B = -\dfrac{1}{4}$.

Therefore, $\int \dfrac{dx}{x^2 + 4x} = \int \dfrac{dx}{4x} - \int \dfrac{dx}{4(x+4)}$

The correct choice is (E).

11. Since the function is increasing the left Riemann sum rectangles will all lie below the graph of the function and therefore their sum is less than the area between the graph and the x-axis, $\int_1^4 f(x)\, dx$. All of the other choices have part or all of their areas above or partly above the graph and will be larger than the left Riemann sum.

The number of subintervals in this situation does not matter. To compare the relative sizes draw one rectangle or trapezoid on the interval $[1,4]$. The relative sizes of these rectangles will be true for any number of rectangles.

The correct choice is (B).

12. Note the cyclic pattern of derivatives of $y = \sin(2x)$:

$y = \sin(2x)$

$y' = 2\cos(2x) = 2^1 \cos(2x)$

$y'' = -4\sin(2x) = -2^2 \sin(2x)$

$y''' = -8\cos(2x) = -2^3 \cos(2x)$

$y^{(4)} = 16\sin(2x) = 2^4 \sin(2x)$

$y^{(5)} = 32\cos(2x) = 2^5 \cos(2x)$

Each time a derivative is taken, the Chain Rule introduces another factor of 2.

Also note that every "even" derivative of $y = \sin(2x)$ contains $\sin(2x)$ as a factor. So the 20th derivative will contain $\sin(2x)$ as a factor, eliminating choices (C), (D), and (E). The 20th derivative is "similar" to the 4th derivative.

The correct choice is (B).

13. Begin by finding where $f'(x) = 3$:
$$f'(x) = 7 \quad 2x$$
$$3 = 7 - 2x$$
$$-4 = -2x$$
$$2 = x$$

Find the point of tangency: $f(2) = 7(2) - 2^2 = 10$. The equation of the line through $(2,10)$ with a slope of 3 is

$$y - 10 = 3(x - 2)$$
$$y - 10 = 3x - 6$$
$$y = 3x + 4$$

The correct choice is (B).

14. Statement I need not be true. Let $f(x) = x^3$ on $[-1,2]$. $f'(0) = 0$ but $f(-1) \neq f(2)$.

Statement II also need not be true. Again, consider $f(x) = x^3$ on $[-1,2]$. $f'(0) = 0$ but $x = 0$ is not a relative extremum (maximum or minimum).

Statement III also need not be true. Consider $f(x) = x^3 + 2x^2$. Although $f'(0) = 0$, $f''(0) \neq 0$.

The correct choice is (A).

15. Multiply the function values at the right side of each interval by the width of the interval. Add these four products to determine the value of the right Riemann sum.

Thus, $0(3 - 1) + 3(7 - 3) + 3(8 - 7) + (-4)(10 - 8) = 7$.

The correct choice is (C).

16. The Taylor series (about $x = 0$) for e^x is $e^x = 1 + x + \frac{x^2}{2!} + \frac{x^3}{3!} + \cdots$

Substituting $(-4x)$ for x, we get $e^{-4x} = 1 + (-4x) + \frac{(-4x)^2}{2!} + \frac{(-4x)^3}{3!} + \cdots$

So $e^{-4x} \approx 1 - 4x + \frac{16x^2}{2} - \frac{64x^3}{6} = 1 - 4x + 8x^2 - \frac{32}{3}x^3$

The correct choice is (D).

17. The graph of a function changes from concave down to concave up where its second derivative changes from negative to positive. This will occur where the first derivative changes from decreasing to increasing. The first derivative changes from decreasing to increasing at its local minimum points. Thus, the graph of g changes from concave down to concave up where g' has local minimums.

The correct choice is (D).

18. Let $u = \ln x \Rightarrow du = \frac{dx}{x}$

$$\int_e^{e^2} \frac{dx}{x \ln x} = \int_{x=e}^{x=e^2} \frac{du}{u} = \ln|u| = \ln|\ln x| \Big|_e^{e^2} = \ln(\ln e^2) - \ln(\ln e) = \ln 2 - \ln 1 = \ln 2$$

Note: $\ln e^2 = 2$, $\ln e = 1$, and $\ln 1 = 0$

The correct choice is (A).

19. $g(x) = f(3x)$, then using the Chain Rule, $g'(x) = f'(3x)(3)$.

Therefore, $g'(0.1) = f'(0.3)(3) = (1.096)(3) = 3.288$

The correct choice is (E).

20. The two circles, $r = 2\cos\theta$ and $r = 2\sin\theta$, intersect at the pole since $\left(0, \frac{\pi}{2}\right)$ satisfies the first equation and $(0,0)$ the second equation. The other intersection point $\left(\sqrt{2}, \frac{\pi}{4}\right)$ is where $2\cos\theta = 2\sin\theta$, or $\sin\theta = \cos\theta$.

Therefore, the area $A = 2\int_0^{\frac{\pi}{4}} \frac{1}{2}(2\sin\theta)^2 \, d\theta = 4\int_0^{\frac{\pi}{4}} \sin^2\theta \, d\theta$

The correct choice is (B).

21. First recognize the limit as the derivative of the function $f(x) = 2x^5 - 5x^3$.

By definition, $f'(x) = \lim\limits_{h \to 0} \dfrac{f(h+h) - f(x)}{h} = \lim\limits_{h \to 0} \dfrac{[2(x+h)^5 - 5(x+h)^3] - [2x^5 - 5x^3]}{h}$

$$= \lim\limits_{h \to 0} \frac{2(x+h)^5 - 5(x+h)^3 - 2x^5 + 5x^3}{h}$$

Therefore the answer is $f'(x) = 10x^4 - 15x^2$.

The correct choice is (D).

22. Using one of the properties of the definite integral, $\displaystyle\int_2^4 f(x)\, dx + \int_4^8 f(x)\, dx = \int_2^8 f(x)\, dx$.

Or, $6 + \displaystyle\int_4^8 f(x)\, dx = -10 \Rightarrow \int_4^8 f(x)\, dx = -16$.

Therefore $\displaystyle\int_8^4 f(x)\, dx = -\int_4^8 f(x)\, dx = -(-16)$ or 16.

Note: $\displaystyle\int_a^b f(x)\, dx = -\int_b^a f(x)\, dx$.

The correct choice is (E).

23. $y' = 3x^2 + 2ax + b$

$y'' = 6x + 2a$

Since $x = 2$ is a point of inflection, $y''(2) = 0$; therefore, $6(2) + 2a = 0 \Rightarrow a = -6$.

Using $a = -6$ in the original equation and using the fact that when $x = 2$, $y = 0$,
$0 = 2^3 - 6(2)^2 + b(2) - 8$.

Therefore solving for b, $b = 12$.

The correct choice is (D).

24. The acceleration vector is the second derivative of the position vector, hence

$$x = 4t^2 \Rightarrow x' = 8t \Rightarrow x'' = 8; \text{ and } y = \sqrt{t} = t^{\frac{1}{2}} \Rightarrow y' = \frac{1}{2}t^{-\frac{1}{2}} \Rightarrow y'' = -\frac{1}{4}t^{-\frac{3}{2}} = \frac{-1}{4t^{\frac{3}{2}}}.$$

Therefore, at $t = 4$, the acceleration vector is $\left\langle 8, -\dfrac{1}{32} \right\rangle$.

The correct choice is (B).

25. Statements I and II must be true by the Extreme Value Theorem.

Statement III is not necessarily true. Consider $f(x) = x^2$ on $[1,4]$. $f'(c) \neq 0$ for $1 < c < 4$.

The correct choice is (C).

26. Using the ratio test, $\displaystyle\lim_{n \to \infty} \left| \frac{u_{n+1}}{u_n} \right| = \lim_{n \to \infty} \left| \frac{x^{n+1}}{n+1} \cdot \frac{n}{x^n} \right| < 1$

$$\Rightarrow \lim_{n \to \infty} \frac{n}{n+1} |x| < 1$$

$$\Rightarrow |x| < 1 \left(\text{since } \lim_{n \to \infty} \frac{n}{n+1} = 1 \right)$$

$$\Rightarrow -1 < x < 1$$

Now test the endpoints.

When $x = 1$, the series is the alternating harmoinc series $1 - \frac{1}{2} + \frac{1}{3} - \frac{1}{4} + \cdots$, which converges.

When $x = -1$, the series is the negative of the harmonic series $-1 - \frac{1}{2} - \frac{1}{3} - \frac{1}{4} + \cdots$, which diverges.

Therefore, the series convergs for $-1 < x \leq 1$.

The correct choice is (C).

27. Since $\cos x = \displaystyle\sum_{n=0}^{\infty} \frac{(-1)^n x^{2n}}{(2n)!}$, the given series is the Taylor series of $\cos x$ with $x = \pi$.

$\cos \pi = -1$.

The correct choice is (B).

28. Solve the differential equation by separating the variables and integrating both sides of the equation:

$y \, dx = x \, dx$

$y^2 = x^2 + C$ Substitute the initial condition $y(3) = 4$ to find C.

$4^2 = 3^2 + C \Rightarrow C = 7$

Therefore, $y^2 = x^2 + 7$ or $x^2 - y^2 = -7$

The correct choice is (A).

29. I is the integral which will give the area using vertical rectangles. II and III are identical except for the "dummy" variable of integration; they both represent the area found using horizontal rectangles.

    The correct choice is (E).

30. The fourth derivative is part of the coefficient of the fourth-degree term.

    $$\frac{f^{(4)}(0)}{4!}x^4 = \frac{x^4}{2(4)}$$

    $$f^{(4)}(0) = \frac{4!}{8} = 3$$

    The correct choice is (C).

31. An analysis of the derivative, $f'(x)$, may be summarized by the number line

    From the First Derivative Test: moving from left to right, f decreases to a minimum at $x = a$, increases to a maximum at $x = b$, decreases to a minimum at $x = c$, and then increases thereafter. This is graph (B).

    The correct choice is (B).

32. This is an initial value problem for which it is difficult or impossible to find an antiderivative. Nevertheless, an expression can be written for the solution. The general form, given $f'(x)$ and $f(c)$, is $f(x) = f(c) + \int_c^x f'(t)\, dt$ by the Fundamental Theorem of Calculus.

    Notice: the lower limit of integration is the x-coordinate of the initial condition. The function f starts at this value and accumulates from there for each value of x. Use your calculator to do the computation.

    $$f(x) = f(3) + \int_3^x \frac{\sin(1+t^2)}{t^3 - 2t}\, dt$$

    $$f(5) = 7 + \int_3^5 \frac{\sin(1+t^2)}{t^3 - 2t}\, dt$$

    $$f(5) \approx 6.992$$

    The correct choice is (D).

33. The given information implies that the series is absolutely convergent, so II is true.

    If a series is absolutely convergent, then it is convergent, so I is true.

    By moving the negative sign in front of $\sum$, then III reduces to I, so III is also true.

    The correct choice is (E).

34. Graph the function in a suitable window such as $x[-2,5]$ by $y[-5,15]$; The graph is tangent to the x-axis at $x = 2$.

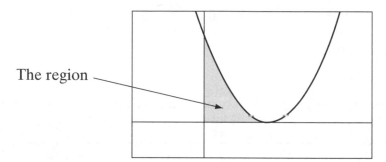

The region

    The side of each square cross section lies in the xy-plane and its length is the y-coordinate of the point on the graph; its area is y^2.

    The thickness of each piece is in the x-direction and is represented by dx.

    The total volume may be found by using the definite integral to sum the individual pieces.

    $\int_0^2 (3(x-2)^2)^2\, dx = 57.6$ using a graphing calculator.

    The correct choice is (E).

35. By the Fundamental Theorem of Calculus, $g' = f$ and $g'' = f'$. Points of inflection of g will occur where its second derivative changes sign. From the table, $f'(x)$ changes sign at $x = 2$ and $x = 6$. Thus g has points of inflection at those two values of x.

    The correct choice is (D).

36. The Lagrange form of the remainder of a Taylor polynomial is given by

$$R_n(x) = \frac{f^{(n+1)}(Z)}{(n+1)!}(x-c)^{n+1}$$ where Z is some number in the interval of convergence.

Substituting the seventh derivative, $c = 0$ and $x = 1$ into this expression gives

$$R_6(1) = \frac{f^7(Z)}{7!}(1-0)^7 = \frac{10,000\cos(Z)}{7!}(1)$$

Since the largest $\cos(Z)$ can be is $+1$, the largest $R_6(1)$ can be is $\frac{10,000}{7!} = 1.984$.

The correct choice is (E).

37. Using the Fundamental Theorem of Calculus,

$$f(1) - f(0) = \int_0^1 \frac{\tan^2 x}{x^2 + 1} dx$$

$$\tfrac{1}{2} - f(0) \approx 0.3446$$

$$-f(0) \approx 0.3446 - 0.500$$

$$f(0) \approx 0.1554$$

There is no easy method to find an antiderivative for the given expression. Therefore you must evaluate the definite integral on your calculator.

The correct choice is (D).

38. $f'(0) > 0$ and $f'(6) > 0$ since the function is increasing for $x \le 8$.

$f''(4) < 0$ since the graph is concave down for $x \le 10$.

$f''(10) = 0$ since this is a point of inflection.

$f''(12) > 0$ since the graph is concave up for $x \ge 10$.

Therefore, the smallest value is $f''(4)$.

The correct choice is (C).

39. Let $p(t) =$ price of the car at time t

$\quad\quad\quad p'(t) =$ rate of change of the price of the car

It is given that $p'(t) = 120 + 180\sqrt{t}$.

Thus $p(t) = \int p'(t)\, dt = \int (120 + 180\sqrt{t})\, dt = 120t + 120t^{\frac{3}{2}} + C$

Since $p(0) = 14,500$, so $C = 14,500$.

Therefore $p(t) = 120t + 120t^{\frac{3}{2}} + 14{,}500 \Rightarrow p(5) = 120(5) + 120(5)^{1.5} + 14{,}500 \approx 16{,}440$.

This may also be approached as an accumulation function.

$$p(5) = 14{,}500 + \int_0^5 (120 + 180\sqrt{t}\,)\, dt \approx 16{,}440$$

The correct choice is (C).

40. The area is given by $\displaystyle\int_0^k (2x - \sin x)\, dx = x^2 + \cos x \Big|_0^k = k^2 + \cos k - 1$

    Solve the equation $k^2 + \cos k - 1 = 0.1$ on your calculator; $k = 0.444$.

    The correct choice is (A).

41. Calculate the change in the value of the derivative from the previous row. The values are shown in the following table:

x	$f'(x)$	$\Delta f'(x)$
0.998	0.980	
0.999	0.995	0.015
1.000	1.000	0.005
1.001	0.995	−0.005
1.002	0.980	−0.015

    The change in the derivative is first positive then negative, indicating that the slope increases then decreases. This indicates a point of inflection in the range covered by the table.

    The correct choice is (C).

42. The fifth week extends from $t = 4$ to $t = 5$. The average rate of change is given by
    $$\frac{m(5) - m(4)}{5 - 4} = \frac{0.92758 - 0.27299}{1} = 0.655$$
    The values are found by evaluating the function with a graphing calculator.

    The correct choice is (D).

43. <u>Method I:</u>

The integrals give the area of the region between the x-axis and the graph of the function. Since the region below the x-axis is counted as negative the greatest value will be $\int_0^2 f(x)\,dx$.

<u>Method II:</u>

The choices may be thought of as various values of the function $F(t) = \int_0^t f(x)\,dx$. Since $F'(t) = f(t)$ by the Fundamental Theorem of Calculus, the largest value of F will occur when its derivative, f, changes from positive to negative. From the graph this occurs only at $t = 2$, so $F(2) = \int_0^2 f(x)\,dx$ will be the absolute maximum.

The correct choice is (B).

44. Since the given limit is in indeterminate form, $\frac{0}{0}$, use L'Hôpital's Rule.

$$\lim_{h \to 0} \frac{g(1+h) - g(1-h)}{h} = \lim_{h \to 0} \frac{g'(1+h) + g'(1-h)}{1} = g'(1) + g'(1) = 2g'(1).$$

Since $g(x) = e^{2x}, g'(x) = 2e^{2x}$ so $g'(1) = 2e^2$ and $2g'(1) = 4e^2$.

The correct choice is (D).

45. The limit is the Riemann sum for the function $x \cos x$ on the closed interval $[0, \pi]$. Therefore,

$$\lim_{n \to \infty} \sum_{i=1}^{n} x_i \cos(x_i)\, \Delta x = \int_0^{\pi} x \cos x\, dx = -2, \text{ using your graphing calculator.}$$

The correct choice is (A).

1a. Speed = $\sqrt{x'(t)^2 + y'(t)^2}$

$\qquad x(t) = 2 + \sqrt{t}\,;\, x'(t) = \dfrac{1}{2\sqrt{t}}$

$\qquad \dfrac{dy}{dt} = y'(t) = te^t - e^t$

Speed at time $t = 3$ is $\sqrt{\left(\dfrac{1}{2\sqrt{3}}\right)^2 + (3e^3 - e^3)^2} = 40.172$

1b. Distance $= \displaystyle\int_0^3 \sqrt{\left(\dfrac{dx}{dt}\right)^2 + \left(\dfrac{dy}{dt}\right)^2}\, dt$

$\qquad\qquad\quad = \displaystyle\int_0^3 \sqrt{\left(\dfrac{1}{4t}\right) + (te^t - e^t)^2}\, dt$

$\qquad\qquad\quad = 24.1644$

1c. Since the slope if 5, so $\dfrac{dy}{dx} = \dfrac{\dfrac{dy}{dt}}{\dfrac{dx}{dt}} = 5.$

$\qquad \dfrac{dy}{dx} = 2\sqrt{t}\,(te^t - e^t)$

Setting $2\sqrt{t}\,(te^t - e^t) = 5$, $t = 1.4725079$

1d. By the Fundamental Theorem of Calculus,

$$\int_0^3 y'(t)\, dt = y(3) - y(0)$$

$$\int_0^3 (te^t - e^t)\, dt = y(3) - (-2)$$

$$22.0855 = y(3) + 2$$

$$y(3) = 20.0855$$

<u>Note:</u> If "integration by parts" was used, then the exact answer would be e^3.

2a. Graph, $A(t)$ and $B(t)$.

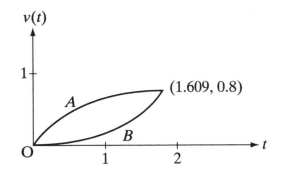

Because of the way the question is worded ("They travel until they have the same velocities."), the graph should end when the graphs intersect.

2b. Person A.

The graphs are velocities, therefore, the area under them represents the distance traveled. Since the area under the graph of A is larger, person A has traveled farther. Alternately, each distance could be calculated (A's distance is $\displaystyle\int_0^{1.609} (1 - e^{-t})\, dt$ see (c) and (d) below) and the answers compared).

2c. $\displaystyle\int_0^x ((1 \quad e^{-t}) - 0.2\,(e^t - 1))\, dt$

If x represents the time traveled, then $\displaystyle\int_0^x ((1 - e^{-t}) - 0.2\,(e^t - 1))\, dt$ represents the distance between the walkers as a function of time.

2d. At $t = 1.609$ minutes

The distance is greatest when the velocities are equal; after this time, B will be walking faster and gaining on A. This occurs when the graphs meet at $t = 1.609$ minutes. At this point, the area under the graph of B begins increasing faster than the area under A.

Alternately, the greatest distance will occur when the derivative of (c), $(1 - e^{-t}) - 0.2\,(e^t - 1)$ is equal to zero. Solving this equation gives $t = 0$, a minimum (when they start), and $t = 1.609$ minutes, a maximum (when B begins to travel faster than A).

Graph $f'(x)$ in the window $x[0,2]$ and $y[-2,2]$.

REMINDER: Numerical and algebraic answers need not be simplified, and if simplified incorrectly credit will be lost.

3a. $g(3) = -3 + \int_2^3 f(t)\, dt = -3 + \frac{1}{2} = -2.5$

$g'(3) = -1 + f(3) = -1 + 1 = 0$

3b. $g'(x) = -1 + f(x) = 0$

$f(x) = 1$

From the graph, you can see that $f(x)$ will equal 1 when $x = 1, 3, 4.5$

3c. Since $g'(x) = -1 + f(x)$, then $g''(x) = f'(x)$.

3d.

x	$g'(x)$	$g''(x)$	Conclusion
1	0	-1	Relative maximum by the Second Derivate Test
3	0	1	Relative minimum by the Second Derivative Test
4.5	0	-2	Relative maximum by the Second Derivative Test

4a. The midpoints of the 5 intervals are $t = 0.1, 0.3, 0.5, 0.7,$ and 0.9. The midpoint sum is

$$(0.2)(90 + 80 + 80 + 40 + 10) = 60$$

4b. The antiderivative of acceleration is the velocity. Thus $f(1)$ is the velocity of the train at the end of the first hour. The units are miles/hour.

4c. $v(t) = \int a(t)\, dt = \int 90\, dt$

$v(t) = 90t + C$

$v(0) = 0 = 90(0) + C$

$\quad 0 = C$

$s(t) = \int_{0.1}^{0.2} v(t)\, dt = \int_{0.1}^{0.2} 90t\, dt$

$s(t) = 1.35$ miles

5a.

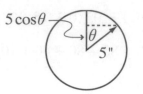

The line indicated in the figure has a length of $5\cos\theta$. Add this to 35 inches that the center is above the floor to get the expression $y = 35 + 5\cos\theta$.

5b. Since the hand travels through 2π radians each 60 minutes, the change in θ is $\frac{d\theta}{dt} = 2\pi \cdot \frac{1}{60} = \frac{\pi}{30}$ radians per minute. By integration, $\theta = \frac{\pi}{30}t$. (Since θ and t both start at zero, the constant of integration is zero.)

5c. Differentiate your answer to (a) with respect to t and substitute for $\frac{d\theta}{dt}$.

$$\frac{dy}{dt} = -5\sin\theta \cdot \frac{d\theta}{dt} = -5\sin\theta \cdot \frac{\pi}{30} = -\frac{\pi}{6}\sin\theta$$

5d. The answer to (c) gives the rate of change in y. To find when the rate of change is largest (when the increase is most rapid), find its derivative $y'' = -\frac{\pi}{6}\cos\theta\frac{d\theta}{dt}$. Since $\frac{d\theta}{dt}$ is constant, y'' will be zero when $\cos\theta = 0$. This occurs when $\theta = \frac{\pi}{2}$ and $\frac{3\pi}{2}$ or 15 and 45 minutes after the hour. The distance is increasing when the hand is moving upwards; this at 45 minutes after the hour.

6a. Separate the variables and solve the differential equation:

$$\frac{dT}{dt} = 14 - 0.01T$$

$$\frac{dT}{14 - 0.01T} = dt$$

$$-100 \ln(14 - 0.01T) = t + k_1$$

$$\ln(14 - 0.01T) = -0.01t + k_2, \; k_2 = \frac{k_1}{-100} = -0.01k_1$$

$$14 - 0.01T = k \cdot e^{-0.01t}, k = e^{k_2}$$

$$T(t) = -100(k \cdot e^{-0.01t} - 14)$$

$$T(t) = 1400 - k \cdot e^{-0.01t}$$

6b. Substitute $t = 0$ and $T = 0$ to find k:

$$0 = 1400 - k \cdot e^0$$

$$0 = 1400 - k(1)$$

$$k = 1400$$

6c. The temperature is given by $T(t) = 1400 - 1400e^{-0.01t}$.

As $t \to \infty$, that is, if the burner is left on a long time, $\lim_{t \to \infty} T(t) = 1400$.

The hottest the burner will ever get is about $1400°$F above room temperature. Since this is less than $1500°$F, the burner is safe.

Note: $\lim_{t \to \infty} (1400 - 1400e^{-0.01t}) = \lim_{t \to \infty} 1400(1 - e^{-0.01t}) = 1400$

Sample Examination VI

1. The segments in the slope field will be horizontal when the numerator $x^2 y + y^2 x = xy(y + x) = 0$. This occurs when $x = 0$, or $y = 0$, or $y = -x$.

   The correct choice is (E).

2. This problem is solved using integration by parts. $u = x$, $du = dx$, $dv = e^x dx$, $v = e^x$.

   $$\int_0^2 xe^x dx = uv - \int v\,du = xe^x - \int e^x dx = xe^x - e^x \Big|_0^2 = (2e^2 - e^2) - (0 - e^0) = e^2 + 1$$

   The correct choice is (C).

3. The y-component of the acceleration vector is y''.

   $$y' = \alpha\beta \cos \beta t$$
   $$y'' = \alpha\beta(-\beta \sin \beta t)$$
   $$= -\beta^2 \alpha \sin \beta t$$
   $$= -\beta^2 y \ (\text{since } y = \alpha \sin \beta t)$$

   The correct choice is (A).

4. If $y = \dfrac{3x + 4}{4x - 3}$, then by the Quotient Rule, $y' \dfrac{3(4x - 3) - 4(3x + 4)}{(4x - 3)^2}$, or $y' = \dfrac{-25}{(4x - 3)^2}$. When $x = 1$, $y' = -25$.

   The equation of the tangent line is $y - 7 = -25(x - 1)$ or $y + 25x = 32$.

   The correct choice is (A).

5. The acceleration of a particle $a(t) = v'(t)$ where $v(t) = x'(t)$.

   $$x(t) = \frac{1}{2} \sin t + \cos(2t)$$
   $$v(t) = x'(t) = \frac{1}{2} \cos t - 2 \sin(2t)$$
   $$a(t) = v'(t) = -\frac{1}{2} \sin t - 4 \cos(2t)$$

   Therefore $a\left(\frac{\pi}{2}\right) = -\frac{1}{2} \sin\left(\frac{\pi}{2}\right) - 4 \cos(\pi) = -\frac{1}{2} - 4(-1) = \frac{7}{2}$

   The correct choice is (E).

100

6. By the Chain Rule $h'(x) = f'(g(x)) g'(x)$. Thus

$$h'(-3) = f'(g(-3)) g'(-3) = f'(1) g'(-3) = \left(-\frac{1}{3}\right)(-1) = +\frac{1}{3}$$

The value of $g(-3)$ is read from the graph of $g(x)$, and the derivatives are the slopes of the lines at $g(-3)$ and $f(1)$.

The correct choice is (D).

7. $\displaystyle\sum_{n=1}^{\infty} \frac{3^{n+1}}{4^n} = \sum_{n=1}^{\infty} 3\left(\frac{3}{4}\right)^n = 3\sum_{n=1}^{\infty} \left(\frac{3}{4}\right)^n.$

The series $\displaystyle\sum_{n=1}^{\infty} \left(\frac{3}{4}\right)^n$ is an infinite geometric series whose ratio, r, is $\frac{3}{4}$.

Since $|r| < 1$, the sum is $\dfrac{a}{1-r}$ or $\dfrac{\frac{3}{4}}{1 - \frac{3}{4}} = 3.$

Therefore, $3\displaystyle\sum_{n=1}^{\infty} \left(\frac{3}{4}\right)^n = 3(3) = 9.$

The correct choice is (C).

8. Let $u = x^2 + 16$, so $du = 2x\,dx \Rightarrow x\,dx = \frac{1}{2}du$

$$\int_0^3 \frac{x}{\sqrt{x^2 + 16}}\,dx = \frac{1}{2}\int \frac{du}{\sqrt{u}} = \frac{1}{2}\int u^{-\frac{1}{2}}\,du = \frac{1}{2}\left(2u^{\frac{1}{2}}\right) = u^{\frac{1}{2}} = (x^2 + 16)^{\frac{1}{2}}$$

Therefore, $(x^2 + 16)^{\frac{1}{2}} = \sqrt{x^2 + 16}\,\Big|_0^3 = \sqrt{25} - \sqrt{16} = 5 - 4 = 1.$

The correct choice is (A).

9. $y = \ln(3x + 5)$

$\dfrac{dy}{dx} = \dfrac{3}{3x + 5}$ $\left(\text{The derivative of } \ln u \text{ is } \dfrac{du}{u}\right)$

$\dfrac{d^2 y}{dx^2} = \dfrac{-9}{(3x + 5)^2}$

The correct choice is (D).

10. Using the method of partial fractions,

$$\frac{1}{x^2 + 5x + 6} = \frac{1}{(x+2)(x+3)} = \frac{A}{x+2} + \frac{B}{x+3}$$

$$1 = A(x+3) + B(x+2)$$

Let $x = -3$. Then $1 = -B \Rightarrow B = -1$

Let $x = -2$. Then $1 = A \Rightarrow A = 1$

$$\int \frac{dx}{x^2 + 5x + 6} = \int \frac{dx}{x+2} - \int \frac{dx}{x+3}$$

$$= \ln|x+2| - \ln|x+3| = \ln\left|\frac{x+2}{x+3}\right|$$

Therefore, $\ln\left|\frac{x+2}{x+3}\right|\Big|_{-1}^{1} = \ln\left(\frac{3}{4}\right) - \ln\left(\frac{1}{2}\right)$

$$= (\ln 3 - \ln 4) - (\ln 1 - \ln 2)$$

$$= \ln 3 - 2\ln 2 - \ln 1 + \ln 2$$

$$= \ln 3 - \ln 2$$

$$= \ln\left(\frac{3}{2}\right)$$

The correct choice is (A).

11. By implicit differentiation, $2yy' - [(2x)(y') + 2y] = 0$

$$2yy' - 2xy' - 2y = 0$$

$$2yy' - 2xy' = 2y$$

$$y'(2y - 2x) = 2y$$

$$y' = \frac{2y}{2y - 2x} = \frac{2y}{2(y-x)}$$

$$y' = \frac{y}{y-x}\Big|_{(2,-3)} = \frac{-3}{-2-2} = \frac{-3}{-5} = \frac{3}{5}$$

The correct choice is (E).

12. The average value of $\sqrt{3x}$ on $[0,9]$ is $\frac{1}{9-0}\int_0^9 \sqrt{3x}\,dx$.

$$\int \sqrt{3x}\,dx = \sqrt{3}\int \sqrt{x}\,dx = \frac{2\sqrt{3}}{3}x^{\frac{3}{2}}.$$

Therefore, the average value is $\frac{1}{9}\left(\frac{2\sqrt{3}}{3}x^{\frac{3}{2}}\right)\Big|_0^9 = 2\sqrt{3}$.

The correct choice is (B).

13. The various values of g are found by calculating the areas of the regions between the graph and the x-axis. Those regions below the axis are counted as negative values.

$$g(-4) = -1.5 = -2 + 0.5$$

$$g(-3) = -2$$

$$g(-1) = 0$$

$$g(0) = 1.5 = 0.5(1 + 2)$$

$$g(3) = 0 = 2 - 2$$

Hence $g(-1) = g(3)$.

The correct choice is (C).

14. $$\int_{-\infty}^{\infty} e^{-|x|} dx = \int_{-\infty}^{0} e^x dx + \int_{0}^{\infty} e^{-x} dx$$

$$\int_{-\infty}^{0} e^x dx = \lim_{t \to -\infty} (e^x) \Big|_{t}^{0} = \lim_{t \to -\infty} (1 - e^t) = 1$$

$$\int_{0}^{\infty} e^{-x} dx = \lim_{t \to \infty} (-e^{-x}) \Big|_{0}^{t} = \lim_{t \to \infty} (1 - e^{-t}) = 1$$

Therefore, $\int_{-\infty}^{\infty} e^{-|x|} dx = 1 + 1$ or 2.

The correct choice is (D).

15. $$y = 6x^2 + \frac{x}{2} + 3 + 6x^{-1}$$

$$y' = 12x + \frac{1}{2} - 6x^{-2}$$

$$y'' = 12 + 12x^{-3}$$

$y'' = 0$ when $12 + 12x^{-3} = 0$ or $12(1 + x^{-3}) = 0 \Rightarrow x = -1$

Also note that y'' does not exist at $x = 0$.

Signs of y'':

$$\xleftarrow{\qquad \overset{+}{\qquad} \Big|_{-1} \overset{-}{\qquad} \Big|_{0} \overset{+}{\qquad}} \rightarrow$$

Therefore, the graph of y is concave down for $-1 < x < 0$, where $y'' < 0$.

The correct choice is (B).

16. Let (x, y) be a point on the parabola $y = 12 - x^2$.

    Then the area of the rectangle, $A = (2x)(12 - x^2) = 24x - 2x^3$.

    $A' = 24 - 6x^2 = -6(x^2 - 4) = -6(x - 2)(x + 2) \Rightarrow A' = 0$ when $x = \pm 2$.

    The maximum occurs when $x = 2$. So the area of the rectangle is $A(2) = 32$.

    The correct choice is (D).

17. The units of an accumulation function are the product of the units of the dependent and independent variables. Thus

$$\frac{\text{liters}}{\text{kilometer}} \cdot \text{kilometers} = \text{liters}$$

    If the function $r(c)$ were graphed, the definite integral would give the area under the graph. The units of the horizontal dimension of the "squares" are those of the independent variable, kilometers. The units of the vertical dimension are those of the dependent variable, liters/kilometer. The area of the squares is the product, namely liters.

    The correct choice is (A).

18. The graph of $f(x)$ has 2 relative maxima and 2 relative minima, which means that $f'(x)$ crosses the x-axis in four points.

    The correct choice is (E).

19. Use implicit differentiation to find the derivative:

$$2x - 2y\frac{dy}{dx} + 1 = 0$$

$$\frac{dy}{dx} = \frac{2x + 1}{2y}$$

    The derivative is undefined and the tangent is vertical when $y = 0$. When $y = 0$

$$\begin{aligned} x^2 - 0 + x &= 2 \\ x^2 + x - 2 &= 0 \\ (x + 2)(x - 1) &= 0 \\ x &= -2, 1 \end{aligned}$$

    So the points are $(-2, 0)$ and $(1, 0)$.

    The correct choice is (D).

20. The area enclosed by the graph and the axis is the distance the object has moved from the starting point. When the velocity is positive (above the axis) the object is moving to the right; when the velocity is negative (below the axis) the object is moving to the left. The object moves to the right from the beginning past A and B to point C when the objects begins moving to the left. Therefore, the object is farthest to the right at C.

    The correct choice is (C).

21. The definition of the derivative of f at $x = 2$ is: $f'(2) = \lim\limits_{x \to 2} \dfrac{f(x) - f(2)}{x - 2}$.

    Therefore, from the given information, $\lim\limits_{x \to 2} \dfrac{f(x)}{x - 2} = \lim\limits_{x \to 2} \dfrac{f(x) - f(2)}{x - 2} \Rightarrow f(2) = 0$, so I is true.

    Since $f'(2) = 0$ and $f(2) = 0$, the graph of $f(x)$ has a horizontal tangent at the point $(2,0)$, so III is true. <u>Note</u>: $f(x)$ will be tangent to the x-axis at $x = 2$.

    If a function is differentiable at a point, it must be continuous there. Therefore, since f is differentiable at $x = 2\,(f'(2) = 0)$, f is continuous at $x = 2$, and II is true.

    The correct choice is (E).

22. The volume of a solid of a known cross section area $A(x)$ from $x = a$ to $x = b$ is $V = \displaystyle\int_a^b A(x)\, dx$.

    Since $4x^2 + y^2 = 1$, then $y = \pm\sqrt{1 - 4x^2}$ and the area of a semicircle is $(1/2)\,\pi y^2$.

    (The ellipse $4x^2 + y^2 = 1$ has x-intercepts at $x - \pm 1/2$).

    The volume is $\displaystyle\int_{-\frac{1}{2}}^{\frac{1}{2}} (1/2)\,\pi y^2\, dx = \int_{-\frac{1}{2}}^{\frac{1}{2}} (1/2)\,\pi (1 - 4x^2)\, dx = \pi \int_0^{\frac{1}{2}} (1 - 4x^2)\, dx$

    $$= \pi \left(x - \frac{4x^3}{3} \right)\Bigg|_0^{\frac{1}{2}}$$

    $$= \frac{\pi}{3}$$

    The correct choice is (C).

23. Use the Product Rule to find $f'(x)$:

$$f'(x) = (e^x)\left(\frac{1}{x}\right) + (\ln x)(e^x)$$

$$= \frac{e^x}{x} + e^x \ln x$$

$$f'(e) = \frac{e^e}{e} + e^e \ln e$$

Since $\ln e = 1$, $f'(e) = \frac{e^e}{e} + e^e$ or $e^{e-1} + e^e$.

The correct choice is (A).

24. $\frac{dy}{dx} = y \cos x \Rightarrow \frac{dy}{y} = \cos x \, dx$

$$\Rightarrow \int \frac{dy}{y} = \int \cos x \, dx$$

$$\Rightarrow \ln|y| = \sin x + C$$

$$\Rightarrow |y| = e^{\sin x + C} = e^C e^{\sin x}$$

$$\Rightarrow y = C_1 e^{\sin x}$$

Since $y = 3$ when $x = 0$, $3 = C_1 e^{\sin x}$

$$3 = C_1 (1)$$

Therefore, $y = e^{\ln 3} \cdot e^{\sin x}$ or $y = 3 e^{\sin x}$

The correct choice is (E).

25. The number of subintervals in this situation does not matter. To compare the relative sizes draw one rectangle or trapezoid on the interval $[a,b]$. The relative sizes of these rectangles will be true for any number of rectangles.

The left Riemann sum rectangle(s) will be larger than any of the others.

The correct choice is (B).

26. $f'(x) = 2(x+2)^3 (x-5) + 3(x+2)^2 (x-5)^2$

$$= (x+2)^2 (x-5)(2(x+2) + 3(x-5))$$

$$= (x+2)^2 (x-5)(5x-11)$$

Since $(x+2)^2 \geq 0$, the derivative changes sign only twice. Therefore, there are only two extreme values.

The correct choice is (C).

27. The limit may be treated as a Riemann sum for some function on an interval. The general form of a Riemann sum is $\sum_{k=1}^{n} f(x_k)\left(\frac{b-a}{n}\right)$ where x_k are values in each subinterval. If $b - a = 3$ then the entire interval is three units wide. The first point in the interval occurs when $k = 1$ and $x_1 = 2 + \frac{3}{n}$ so this is a right hand sum; the final value is $x_n = \left(2 + \frac{3}{n}\right) n = 5$ and the function is $f(x) = x^2$.

The limit is then equal to $\int_2^5 x^2 \, dx = \frac{x^3}{3}\Big|_2^5 = \frac{1}{3}(5^3 - 2^3) = 39$.

Note: There are always more than one solution to this type problem, all of which give the same answer. Another choice is $\int_0^3 (2 + x)^2 dx = 39$

The correct choice is (D).

28. Statement I is true – Euler's method uses the point on the tangent line with the same x-coordinate to approximate the value of a function.

Statements II and III are false – Euler's method is not used for finding roots. While Euler's method uses the y-coordinate of B to approximate the y-coordinate of D, there is no way of knowing in advance that the y-coordinate of D is zero.

The correct choice is (A).

29. Let $u = 2x$, so $du = 2\,dx \Rightarrow dx = \frac{1}{2}du$, when $x = 2, u = 4$ and when $x = 3, u = 6$.

$\int_2^3 f(2x) \, dx = \frac{1}{2}\int_4^6 f(u) \, dx \Rightarrow \int_4^6 f(u) \, du - 2\int_2^3 f(x) \, dx$

Therefore, $\int_4^6 f(x) \, dx = 2(8) = 16$

The correct choice is (D).

30. Area, $A = \frac{1}{2}\int_0^\pi (6 \cos \theta + 8 \sin \theta)^2 d\theta = 25\pi \approx 78.540$.

This could be done by paper and pencil, but the better approach is with the use of a calculator. The equation is a circle with center at $(3,4)$ and radius 5.

The correct choice is (C).

31. The volume may be found by multiplying the cross section area, $\int_{-20}^{20} \left(20 - \frac{x^6}{3,200,000}\right) dx$ square feet, by the base (50 feet). It may also be found by forming a Riemann sum of the volumes of slices perpendicular to the x-axis. The base is 50, the height is $y(x)$ and the thickness is dx.

In both cases the volume is given by $50 \int_{-20}^{20} \left(20 - \frac{x^6}{3,200,000}\right) dx \approx 34,300$ cubic feet.

The correct choice is (D).

32. Find where the walls hit the ground by solving

$$20 - \frac{x^6}{3,200,000} = 0$$
$$x^6 = 64,000,000$$
$$x = \pm 20$$

Then the average height is given by $\frac{1}{20-(-20)} \int_{-20}^{20} \left(20 - \frac{x^6}{3,200,000}\right) dx \approx 17.143$ feet, by calculator.

The correct choice is (E).

33. The mass of a piece of the rod of length Δx meters is found by multiplying $\rho(x)$ by Δx. The mass of these pieces are added giving $\sum_{i=1}^{n} \rho(x_i) \Delta x$. The limit of this Riemann sum as $n \to \infty$ is given by $\int_0^1 \rho(x)\, dx = \int_0^1 (1 + (1-x)^2)\, dx = \frac{4}{3} = 1.333$

The units are $\frac{\text{grams}}{\text{meters}} \cdot \text{meters} = \text{grams}$

The correct choice is (B).

34. The area of the shaded region is bounded by the x-axis and $y = f(x)$.

    Since $y = 0$ lies above the curve $y = f(x)$ on $[a,b]$, the area is

    $$\int_a^b [0 - f(x)]\, dx = \int_a^b -f(x)\, dx = -\int_a^b -f(x)\, dx \text{ or } \int_b^a f(x)\, dx.$$

    Keep in mind, that $\int_a^b f(x)\, dx = -\int_b^a f(x)\, dx.$

    The correct choice is (B).

35. Refering to the accompanying diagram,

    $$\tan\theta = \frac{x}{75,000} \Rightarrow \theta = \tan^{-1}\left(\frac{x}{75,000}\right).$$

    Therefore, $\dfrac{d\theta}{dt} = \dfrac{(1/75,000)(dx/dt)}{1+(x/75,000)^2}$

    Since $\dfrac{dx}{dt} = 16,500$ when $x = 38,000$

    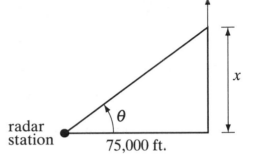
    radar station — θ — 75,000 ft. — x

    $$\frac{d\theta}{dt} = \frac{16,500/75,000}{1+(38,000/75,000)^2} = \frac{165/750}{1+(38/75)^2} \approx 0.175$$

    The correct choice is (A).

36. The error in a Taylor Series is given by the expression $\left|\dfrac{f^{(n+1)}(z)}{(n+1)!}(x-a)^{n+1}\right|$ where z is a number between x and a. For the numbers given, the derivative is always less than one; then with $x = 1.2$ the error is always less than

    $$\left|\frac{1}{(n+1)!}(1.2-1)^{n+1}\right|$$

    Evaluate this expression on your calculator for the answer choices to determine which is less than 0.00001.

    For $n = 3$, the value is 0.0000667; and for $n = 4$, the value is 0.00000267. Therefore, five terms (those with $n = 0, 1, 2, 3$ and 4) are sufficient to approximate the function to the desired accuracy.

    The correct choice is (C).

37. By the Integral Test $\int_1^\infty \frac{1}{x^{\pi p}} \, dx$ will converge if the series $\sum_{n=1}^\infty \frac{1}{n^{\pi p}}$ converges. This series is a p-series and will converge if the exponent $\pi p > 1$. Thus, the series and the integral will converge when $p > \frac{1}{\pi}$.

   The correct choice is (B).

38. The graph of a function will be concave up when $f''(x) > 0$. Since $f''(x) > 0$ for $x < 0$ and $b < x < c$, so I is true.

   The (relative) maximum and minimum values of a function cannot be determined from the second derivative alone, so II is false.

   The function f has points of inflection where $f''(x)$ changes sign. This occurs at $x = 0$ and $x = b$. So $x = 0$ and $x = b$ are points of inflection, so III is true.

   The correct choice is (D).

39. The relative maximum values of a function occur where the first derivative of the function changes from positive to negative. By the Fundamental Theorem of Calculus the derivative is

$$f'(x) = \sin\left(\frac{1}{x}\right)$$

   Graph the derivative in a convenient window such as x [0.1, 0.4] by y [-1.3, 1.3]

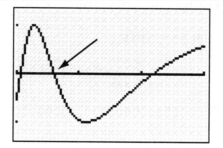

   Using a calculator find the x-coordinate of where the derivative changes from positive to negative (indicated by the arrow).

   The value is $x = 0.159$.

   The other two values, $x = 0.106$ and $x = 0.318$, are the locations of the relative minimums of the function.

   The correct choice is (B).

40. The term with $k = 10$ of any Taylor series is $f^{(10)}(a)\dfrac{(x-a)^{10}}{10!}$.

    For this series, the corresponding term is $\left(\dfrac{1}{2}\right)^{10}\dfrac{(x-3)^{10}}{10!}$

    Therefore, $f^{(10)}(3) = \left(\dfrac{1}{2}\right)^{10} = 9.766 \times 10^{-4}$.

    The correct choice is (C).

41. By the Fundamental Theorem of Calculus (version II),

    $$\frac{d}{dx}\int_0^{2x}(e^t + 2t)\,dt = 2\left(e^{2x} + 2(2x)\right) - 0 = 2e^{2x} + 8x.$$

    The correct choice is (E).

42. By the Mean Value Theorem

    $$f'(c) = \frac{f(b) - f(0)}{b - 0}$$

    $$\cos(4.522) = \frac{\sin b}{b}$$

    Solving this equation with a calculator gives $b \approx 4$.

    The correct choice is (A).

43. The fourth-degree Taylor polynomial for $\cos x$ about $x = 0$ is $1 - \dfrac{x^2}{2!} + \dfrac{x^4}{4!}$.

    Substituting $x = \dfrac{1}{2}$ gives $1 - \dfrac{\left(\frac{1}{2}\right)^2}{2!} + \dfrac{\left(\frac{1}{2}\right)^4}{4!} = 1 - \dfrac{1}{8} + \dfrac{1}{384}$.

    The correct choice is (E).

44. $\displaystyle\int_0^{1000} 8^x\,dx - \int_a^{1000} 8^x\,dx = \int_0^{1000} 8^x\,dx + \int_{1000}^a 8^x\,dx = \int_0^a 8^x\,dx = \left.\dfrac{8^x}{\ln 8}\right|_0^a = \dfrac{8^a - 1}{\ln 8}$

    Solving $\dfrac{8^a - 1}{\ln 8} = 10.40$ on a graphing calculator results in $a = 1.5$.

    The correct choice is (B).

45. The average speed is given by $\dfrac{1}{32,000}\displaystyle\int_0^{32,000} s(a)\,da$. While it is possible to write a piecewise function for $s(a)$ it is easier to find the value of the integral by find the area under the graph directly.

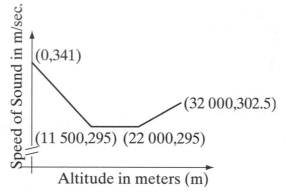

Region A is a trapezoid. The area is $\dfrac{1}{2}(11,500)(341+295)=3,657,000$

Region B is a rectangle. The area is $295(10,5000)=3,097,500$

Region C is a trapezoid. The area is $\dfrac{1}{2}(10,000)(302.5+295)=2,987,500$

The total area is 9,742,000

Divide by 32,000 to find the average value $= 304.4$ m/sec.

Note: It is incorrect to find the average speed on each part and then average the averages. This gives answer (B) 303.9.

The correct choice is (C).

1a.

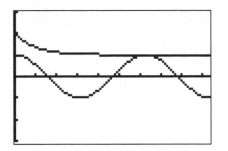

1b. Since for all x, $1 + e^{-x} > 1$ and $\cos x \leq 1$, the particles can never be at a point with the same y-coordinate, therefore they will never collide.

1c. The distance between the particles is $s(x) = 1 + e^{-x} - \cos x$. Differentiate to find the critical values:

$$\frac{ds}{dt} = -e^{(-x)} + \sin x$$

Set the derivative equal to zero and solve for x on a calculator. $x = 0.5885$, $x = 3.0963$, $x = 6.2850$, and $x = 9.4246$

Then $s(0.5885) = 0.7233$; $s(3.0963) = 2.04419$; $s(6.2850) = 0.00186$; $s(9.4246) = 2.0000806$

Now check the endpoints $s(0) = 1$ and $s(3\pi) = 2.0000807$

The minimum distance is about 0.00186 or 0.001 or 0.002

1d. From part (c) the maximum distance is 2.04419 or 2.044

2a. Separate the variables, integrate and solve for R:

$$\frac{dR}{R} = k \cdot dt$$

$$\ln R = k \cdot t + C_1$$

$$R = C \cdot e^{kt}$$

Substitute $t = 0$ and $R = 10^6$ and find C: $10^6 = Ce^0$ and $C = 10^6$

Substitute $t = 100$ and $R = 10^2$ to find k:

$$10^2 = 10^6 e^{100k}$$

$$10^{-4} = e^{100k}$$

$$\ln 10^{-4} = 100k \text{ and } k = \frac{\ln 10^{-4}}{100} \approx -0.092$$

Therefore, $R(t) = 10^6 e^{-0.092t}$

2b. Solve $10 = 10^6 e^{-0.092t}$ for t on a calculator or by hand:

$$10^{-5} = e^{-0.092t}$$

$$\ln 10^{-5} = -0.092t$$

$$t = \frac{\ln 10^{-5}}{-0.092} = 125.140 \text{ or about } 125 \text{ seconds}$$

2c. Solve $\frac{1}{2}10^6 = 10^6 e^{-0.092t}$

$$\frac{1}{2} = e^{-0.92t}$$

$$\ln \frac{1}{2} = -0.092t$$

$$t = \frac{\ln \frac{1}{2}}{-0.092} = 7.534 \text{ seconds}$$

REMINDER: Numerical and algebraic answers need not be simplified, and if simplified incorrectly credit will be lost.

3a. Area $\int_0^1 (e^x - e^{-x})\, dx = e^x + e^{-x}\Big|_0^1$

$$= e + e^{-1} - (e^0 + e^0)$$
$$= e + e^{-1} - 2$$

3b. Volume $= \int_0^1 \pi((e^x)^2 - \pi(e^{-x})^2 dx) = \pi\left(\dfrac{e^{2x}}{2} - \dfrac{e^{-2x}}{-2}\right)\Big|_0^1$

$$= \pi\left(\dfrac{e^2}{2} + \dfrac{e^{-2}}{2} - \left(\dfrac{1}{2} + \dfrac{1}{2}\right)\right) = \pi\left(\dfrac{e^2 + e^{-2} - 2}{2}\right)$$

3c. The length of the base of each rectangle is $(e^x - e^{-x})$ and the cross-section area is $x(e^x - e^{-x})$.

Thus the Volume $= \int_0^1 x(e^x - e^{-x})\, dx$

4a. Integrate the given equations:

$$x(t) = 4\sin t + C$$

$$y(t) = -\cos t + K$$

The constants C and K are found by using the fact that at $t = 0$ both $x = 0$ and $y = 0$.

So $x(0) = 4\sin 0 + C = 0 \Rightarrow C = 0$

and $y(0) = -\cos 0 + K = 0 \Rightarrow K = 1$.

Therefore, $x(t) = 4\sin t$ and $y(t) = -\cos t + 1$.

4b. When $t = 0$, the particle is at $(0,0)$, and at $t = \frac{\pi}{2}$, the particle is at $(4,1)$.

The average rate of change of y with respect to x is $\frac{\Delta y}{\Delta x} = \frac{1-0}{4-0} = \frac{1}{4}$

4c. Set the instantaneous rate of change, $\dfrac{dy}{dx} = \dfrac{dy/dt}{dx/dt} = \dfrac{\sin(t)}{4\cos(t)} = \dfrac{1}{4}\tan(t)$, equal to the average rate of change found in part (b) and solve for t.

$$\frac{1}{4}\tan t = \frac{1}{4}$$

$$\tan t = 1 \Rightarrow t = \frac{\pi}{4}$$

5a. Since $\dfrac{d}{dx}\tan(x) = \sec^2(x)$ the polynomial may be found by differentiating $M(x)$:

$$\frac{d}{dx}M(x) = 1 + x^2 + \frac{2}{3}x^4$$

5b. The 4th degree Maclaurin polynomial for $\dfrac{\tan(x)}{x}$ is $\dfrac{x + \frac{1}{3}x^3 + \frac{2}{15}x^5}{x} = 1 + \frac{1}{3}x^2 + \frac{2}{15}x^4$

5c. $\displaystyle\lim_{x\to 0}\frac{\tan(x)}{x} = \lim_{x\to 0}\left(1 + \frac{1}{3}x^2 + \frac{2}{15}x^4\right) = 1$

5d. The answer is exact.

The Maclaurin polynomial found in (b) is equal to the function it approximates at the center of the interval of convergence (in this case at $x = 0$). Even though the other terms of the power series are not known, they all will contain powers of x. These other terms will all approach zero at $x = 0$ and will not affect the limit.

6a.

6b. $y = xe^{-x} + Ce^{-x}$

$$\frac{dy}{dx} = -xe^{-x} + e^{-x} - Ce^{-x}$$

$$y + \frac{dy}{dx} = (xe^{-x} + Ce^{-x}) + (-xe^{-x} + e^{-x} - Ce^{-x})$$

$$y + \frac{dy}{dx} = e^{-x}$$

6c. $\frac{dy}{dx} = e^{-x}(-x + 1 - C)$

$\frac{dy}{dx} = 0$ when $x = 1 - C$

Substituting $(1 - C)$ for x in the solution equation gives

$y = (1 - C)e^{-(1-C)} + Ce^{-(1-C)} = e^{-(1-C)}$

The maximum points are $(1 - C, e^{-(1-C)})$. They all lie on the graph of $f(x) = e^{-x}$.